Lecture Notes in Physics

Edited by H. Araki, Kyoto, J. Ehlers, München, K. Hepp, Zürich
R. Kippenhahn, München, H.A. Weidenmüller, Heidelberg
J. Wess, Karlsruhe and J. Zittartz, Köln
Managing Editor: W. Beiglböck

299

J. D. Buckmaster T. Takeno (Eds.)

Mathematical Modeling in Combustion Science

Proceedings of a Conference Held in
Juneau, Alaska, August 17–21, 1987

Springer-Verlag
Berlin Heidelberg GmbH

Editors

John D. Buckmaster
Department of Aeronautical and Astronautical Engineering, University of Illinois
101 Transportation Building, 104 South Mathews Avenue, Urbana, IL 61801, USA

Tadao Takeno
Institute of Interdisciplinary Research, The University of Tokyo, Komaba
Meguro-ku, Tokyo 153, Japan

ISBN 978-3-662-13671-3

Library of Congress Cataloging-in-Publication Data. Mathematical modeling in combustion
science. (Lecture notes in physics; 299) 1. Combustion—Mathematical models—Congresses.
I. Buckmaster, John David. II. Takeno, T. (Tadao), 1937-. III. Title. IV. Series.
QD516.M394 1988 541.3'61'0724 88-6666
ISBN 978-3-662-13671-3   ISBN 978-3-540-39131-9 (eBook)
DOI 10.1007/978-3-540-39131-9

2158/3140-543210

PREFACE

In recent years there have been important advances in the mathematics
of combustion, and the subject has attracted the attention of a
significant number of applied mathematicians in the United States.
Useful theoretical contributions are only possible with judicious
modeling, and well-defined, simple experiments can be both stimuli
and guides in this endeavor. Japanese work often shows a clear
recognition of this. For these reasons we thought that it would be
worthwhile to take advantages of the US/Japan Cooperative Science
Program, jointly funded by the National Science Foundation and the
Japanese Society for the Promotion of Science, and organize a
workshop on 'Mathematical Modeling in Combustion Science' which would
bring together a small number of Japanese and American combustion
scientists.

Maximum participation from both sides of the Pacific required a
location in either Hawaii or Alaska. The conventional choice would
have been Hawaii, but we chose Juneau, Alaska, and we strongly
recommend it to any reader contemplating a similar program. Juneau is
a small, friendly town in a striking setting, with several fine
restaurants and the famous Red Dog Saloon, ideal for cross-culutural
socializing. Local attractions for an afternoon break include the
Mendenhall Glacier, salmon spawning streams, and many excellent hiking
trails. Glacier Bay National Park is close-by for the participant
seeking a stunning Alaskan experience during one of the weekends
contiguous with a workshop. In addition, the Baranof hotel has ideal
facilities for a small meeting of twenty or so participants.

Each participant was required to give a single formal lecture;
written versions of these talks are included in these Proceedings.
In addition there were a number of workshop sessions of an informal
nature. Participants had been asked to prepare material in a number
of areas (high Mach number combustion, the interaction between theory
and experiment, small scale turbulence in premixed flames) and were
called upon to provide introductions from which discussions could
then follow. It has not been possible to reproduce those discussions
here, but the introductory material is included. Perhaps it will
stimulate the reader to creative thought as much as it did the
participants.

> J. Buckmaster, Urbana, Il., USA
> T. Takeno, Tokyo, Japan.

TABLE OF CONTENTS

4. DISCUSSION SESSIONS

1. High Mach Number Combustion

SHOCK-INITIATION OF A PLANE DETONATION WAVE

A. K. Kapila
Department of Mathematical Sciences, Rensselaer Polytechnic Institute
Troy, New York 12180-3590

Introduction.

This paper gives a mathematical description of the early stages of evolution of a planar detonation wave, initiated by the passage of a strong shock. It is assumed that the reactive gas undergoes a one-step, first-order, irreversible decomposition reaction governed by Arrhenius kinetics. The analysis is asymptotic, in the limit of large activation energy. There is a deliberate attempt at brevity, since the following presentation draws heavily upon the study reported in [1], to which the reader is referred for further details.

The basis configuration is as follows. For time $t < 0$, the half space $x > 0$ is filled with a reactive gas at a uniform state of rest, and at temperature low enough for the chemical reaction rate to be negligible over any time scales of interest. At $t = 0+$, a piston initially at $x = 0$ is pushed into the gas at a constant speed, thereby generating a shock wave running ahead of it. If the gas were inert, the shock would propagate steadily and maintain a fixed strength. It is assumed, however, that the shock switches on a significant amount of chemical activity in the gas behind it, which then has the effect of strengthening and accelerating the shock. The aim of the following analysis is to describe the sequence of events in the shocked gas until a detonation is about to form.

Governing Equations.

The relevant equations are the reactive Euler equations, which for planar, one-dimensional motion are:

$$(1) \qquad \rho_t + u\,\rho_x + \rho\,u_x = 0,$$

$$(2) \qquad \rho(u_t + uu_x) + \frac{1}{\gamma}\,p_x = 0,$$

(3) $\rho(T_t + u\ T_x) - \frac{\gamma-1}{\gamma}\ (p_t + u\ p_x) = \beta w$,

(4) $\rho(Y_t + u\ Y_x) = -w$,

(5) $p = \rho T$,

(6) $w = \frac{1}{\beta\theta}\ \rho Y\ \exp\left[\theta - \frac{\theta}{T}\right]$.

Here p, ρ, T, u, and Y are, respectively, the gas pressure, density, temperature, velocity and reactant mass fraction. The reference coordinate frame has been selected to move with the piston face, and the reference state of the gas is taken to be the shocked state at t = 0+. Velocity is referred to the acoustic speed and time to the induction time at the reference state; their product then defines the reference length. The dimensionless parameters appearing above are the dimensionless chemical heat release β, the specific heats ratio γ and the dimensionless activation temperature θ.

The appropriate boundary conditions for the shocked region under study are

$$u = 0 \quad \text{at} \quad x = 0,$$

and the Rankine-Hugoniot jump conditions for T, p, u and ρ at the shock locus $x_s(t)$. Immediately behind the shock, Y has the value unity.

In the following analysis, β and γ are assumed fixed and O(1), and the asymptotic limit θ → ∞ is employed. The resulting sequence of events proceeds as follows:

Induction State

Initially, T-1 = $O(\theta^{-1})$ is an appropriate range to consider, as the reaction-rate expression in (6) suggests. Accordingly, one sets

$$u = \theta^{-1}u_1 + \ldots; \quad \phi = 1 + \theta^{-1}\phi_1 + \ldots \quad \text{for} \quad \phi = T, p, \rho \text{ and } Y,$$

to obtain leading-order reduced equations

(7) $\left(\frac{\partial}{\partial t} \pm \frac{\partial}{\partial x}\right)(p_1 \pm \gamma\ u_1) = \gamma\ \exp(T_1)$,

(8) $\frac{\partial}{\partial t}\ (T_1 - \frac{\gamma-1}{\gamma}\ p_1) = \exp(T_1)$,

(9) $\rho_1 = p_1 - T_1$, $\frac{\partial y_1}{\partial t} = -\frac{1}{\beta}\ \exp(T_1)$.

These equations are subject to

$$u_1(0,t) = 0$$

and the linearized R-H conditions at

$$x = Mt,$$

the undisturbed shock locus, where M is the initial shock speed. The above problem can be solved numerically, first for T_1, p_1 and u_1 and then for ρ_1 and Y_1
(see, e.g., [2] or [3] for details). The solution exhibits thermal runaway, which
is characterized by the appearance of logarithmic singularities in T_1 and p_1 (and
therefore Y_1), first at the piston face, at a definite time $t_e(M)$. It is note-
worthy that ρ_1 remains bounded. Typical profiles of $T_1(x,t)$ for t smaller than,
but close to, t_e are shown in Figure 1. These profiles display the emergence of a

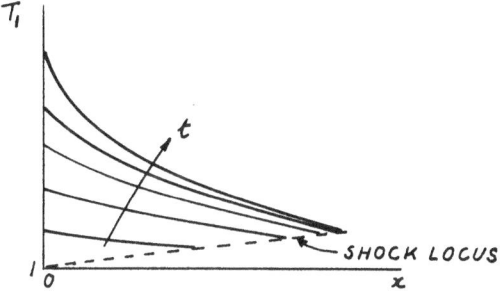

Figure 1

shrinking boundary layer near x = 0 within which the solution grows rapidly. One
can, in fact, continue the induction solution beyond $t > t_e$; the locus of infini-
ties in the solution then moves into the interior of the domain (Figure 2) at a
speed which is initially supersonic, but fails monotonically to the sonic value.
More about this locus will be said later.

It is possible to give an analytical description of the boundary layer at the
piston face in the limit $\tau \to 0+$, where

$$\tau = t_e - t.$$

The boundary layer is found to be $O(\tau^{\gamma/(2\gamma-1)})$ thick, and therefore, describable in terms of the spatial coordinate ξ, defined by

$$x = \xi \, \tau^{\gamma/(2\gamma-1)}.$$

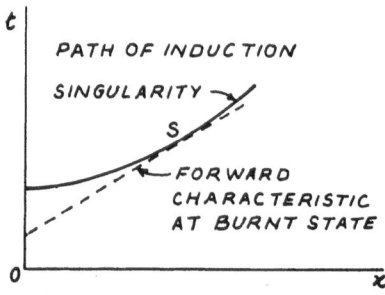

Figure 2

The boundary-layer solution is found to have the form

$$T_1 \sim - \ln(\gamma\tau) + F_0(\xi) + \ldots \, ,$$

$$P_1 \sim - \ln(B_1\tau) + F_0(\xi) + \ldots \, ,$$

$$u_1 \sim \tau^{(\gamma-1)/(2\gamma-1)}[\ln\tau \, H_0(\xi) + H_1(\xi)] + \ldots \, ,$$

where

$$F_0(\xi) = - \ln\left[1 + \alpha A_1 \, \xi^{(2\gamma-1)/\gamma}\right] \, ,$$

$$H_0(\xi) = - \frac{\gamma-1}{\gamma^2} \alpha A_1 \, \xi^{(\gamma-1)/\gamma} \, ,$$

$$H_1(\xi) = \frac{2\gamma-1}{\gamma^2} \alpha A_1 \, \xi^{(\gamma-1)/\gamma} \left[\ln \xi + F_0(\xi) + B_2\right] \, ,$$

and the constants α, A_1, B_1 and B_2 are known. Observe that to leading order the boundary layer displays a spatially uniform growth of temperature and pressure, with spatial structure appearing only as a perturbation. The structure is singular at $\xi = 0$, but this singularity can be removed by means of a thinner, inner layer in which $x = O(\tau)$; details can be found in [1]..

Explosion Stage

The layer solution becomes nonuniform when $-\ell n(\tau)$ becomes $O(\theta)$, suggesting that further evolution should occur on the time scale σ, defined by

$$\tau = e^{-\theta\sigma}, \quad \sigma > 0 .$$

The solution now turns out to have the expansions

$$(10) \quad T \sim T_0(\sigma) + \theta^{-1} \tilde{T}_1(\xi,\sigma) + \cdots ,$$

$$(11) \quad p \sim T_0(\sigma) + \theta^{-1} \tilde{p}_1(\xi,\sigma) + \cdots ,$$

$$(12) \quad u = O[e^{-\theta\sigma(\gamma-1)/(2\gamma-1)}]$$

$$(13) \quad Y \sim Y_0(\sigma) + \theta^{-1} \tilde{Y}_1(\xi,\sigma) + \cdots ,$$

$$(14) \quad \rho \sim 1 + \theta^{-1} \rho_{01} + \cdots ,$$

where

$$(15) \quad T_0(\sigma) = \frac{1}{1-\sigma} , \quad Y_0(\sigma) = \frac{1+\beta\gamma - T_0(\sigma)}{\beta\gamma} ,$$

$$(16) \quad \tilde{T}_1 = - T_0^2 \, \ell n[\gamma \, Y_0/T_0^2] - T_0^2 \, \ell n \left[1 + \frac{\alpha A_1}{T_0^2} \xi^{(2\gamma-1)|\gamma}\right], \quad \tilde{p}_1 = \tilde{T}_1 + T_0 \rho_{01} ,$$

$$(17) \quad \tilde{Y}_1 = - \frac{1}{\beta\gamma} \tilde{T}_1 + B_3 ,$$

and ρ_{01} and B_3 are known constants. Observe that density is essentially unchanged from its value at runaway, i.e., the material within the layer is inertially confined. As σ increases, T and p increase whereas Y decreases. Eventually, p and T peak when Y_0, the leading term in Y, vanishes. This happens at

$$\sigma = \beta\gamma/(1+\beta\gamma) ,$$

and the peak values are

$$T \sim 1+\beta\gamma, \quad p \sim 1+\beta\gamma .$$

At the same time, the $O(\theta^{-1})$ term (in T, say; see (16)) develops a logarithmic singularity, indicating the breakdown of the solution.

Before advancing further in time, it is instructive to point out that as the boundary layer recedes towards the piston during the explosion stage, it leaves behind it an exponentially thin intermediate region in which the solution is essentially stationary in time, but is not close to the induction solution. This region is governed by the spatial variable X, defined by

$$X = e_{-\theta x} \quad, \quad 0 < X < \sigma\gamma/(2\gamma-1) \;,$$

where the left restriction on X corresponds to the edge of the boundary layer and

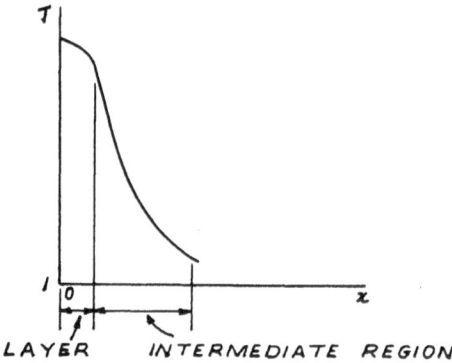

LAYER   INTERMEDIATE REGION

Figure 3

the right restriction to merging with the outer region (see Figure 3). The solution in the intermediate region is given by

$$T \sim \frac{1}{1 - \frac{2\gamma-1}{\gamma} X} \;,$$

with analogous expressions for p and Y, while

$$\rho \sim 1 \;, \quad u \sim \frac{\gamma-1}{\gamma^3} (2\gamma-1)\alpha \, A_1 X \;.$$

Transition stage

Further evolution of the solution near the piston face occurs on the time scale ϕ, defined by

$$t = t_e + v(\theta)\phi \;,$$

where

$$v = \beta\theta \exp[-\beta\gamma\theta/\{1+\beta\gamma\}] \;.$$

The corresponding spatial variable in the boundary layer (in view of its $O(\tau^{\gamma/(2\gamma-1)})$ thickness), is z, defined by

$$x = v^{\gamma/(2\gamma-1)} z \;.$$

The transition zone analysis is rather involved, but the asymptotic form of its solution for large ϕ and z is particularly simple and of special interest. One finds that

$$T \sim 1+\beta\gamma - \theta^{-1}\ \overline{T}_1$$

in this zone, and that

$$\overline{T} \sim \lambda_1 \exp[-\phi + \overline{A}\ z^{(2\gamma-1)/\gamma}] + \text{constant as } \phi, z \to \infty ,$$

where λ and $\overline{A}$ are known constants. Thus, a reaction wave is born, propagating out of the transition zone with velocity

$$dx/dt \sim v^{-(\gamma-1)/(2\gamma-1)} dz/d\phi \sim x^{-(\gamma-1)/\gamma} .$$

Behind the wave the gas is completely reacted, and pressure and temperature are at peak values, $1+\beta\gamma$ to leading order, while density has the leading-order value of unity and velocity is exponentially small. As time continues to evolve, this wave sweeps across the intermediate zone via a sequence of Semenov explosions. Particle velocities, and hence density changes, continue to be exponentially small during this process; the wave is completely reactive in character. In fact, it can be thought of as a nearly-constant-density detonation wave, travelling at velocity supersonic relative to the burnt gas behind it.

As the wave propagates across the intermediate zone, its velocity decreases because of the falling temperature immediately ahead of it, and by the time the wave reaches the edge of the intermediate zone, its velocity has fallen to within a few multiples of the sound speed in the burnt gas. Local chemical times have now risen sufficiently to be comparable to the local acoustic times, and the gas in the vicinity of the wave is no longer inertially confined. The wave path must now be computed numerically, and is in fact given to a first approximation by the thermal runaway locus computed by the induction zone analysis (Figure 2). Even the numerical description is valid only so long as the wave remains supersonic, breaking down at point S in Figure 2 where the computed wave path is tangential to the forward characteristic originating in the burnt gas. At this point, shock formation, and the consequent birth of a conventional detonation, is imminent. A full description of these processes will be given elsewhere.

Acknowledgements

This paper has benefited from discussions with J. W. Dold. The research reported here was supported by the U. S. Army Research Office and by the Los Alamos National Laboratories.

References

1. T. L. Jackson, A. K. Kapila and D. S. Stewart, Evolution of a reaction center in an explosive material, SIAM J. Appl. Math. Submitted for publication (1987).

2. T. L. Jackson and A. K. Kapila, Shock-induced thermal runaway, SIAM J. Appl. Math., 45, 130 (1985).

3. J. F. Clarke and R. S. Cant, Nonsteady gasdynamic effects in the induction domain behind a strong shock wave, Progress in Astro. and Aero., 95, 142 (1984).

EFFECTS OF PREIGNITION FLUCTUATION GROWTH
ON REDUCING THE INDUCTION PERIOD

Shunichi Tsugé and Hiroshi Kohmoto
School of Engineering Sciences, University of Tsukuba
Tsukuba, Ibaraki 305 Japan

ABSTRACT

Growth of fluctuation during preignition peoriod, typically observed for
hydrogen-oxygen premixed gases to occur at about 1000 K, is analyzed in
some detail. The temperature fluctuation growth preceding ignition
affects little on induction time at higher temperatures, whereas at lower
temperatures corresponding to the mild ignition, the value is reduced
considerably, in better agreement with observed data as compared with
the previous ones based on a crude model with constant fluctuation.

I. INTRODUCTION

It has long been puzzled since finding by Voevodski ans Soloukhin[1]
that hydrogen-oxygen premixed gases change manners of ignition sharply
in a narrow range of the ground temperature of about 1000 K. The observation shows that at higher temperatures ignition takes place at a planer
front, which developes into one-dimensional acoustic waves (the strong
ignition). At lower temperatures, on the other hand, ignition starts
at random, spotty reaction centers (the mild ignition). The stochastic
structure in the latter case is shown to enhance the overall reaction
rate enormously, resulting in reduced induction time by the factor of
$10^1 \sim 10^2$ as compared with the classical prediction.[1]

This discrepancy is bridged by using turbulent reaction rate formula[2]
derived on the basis of nonequilibrium statistical mechanics. The formula
accounts for the reaction rate elevation far beyond the first order small-
ness of the relative temperature fluctuation if the energy of activation
is sufficiently high. This phenomenon may be regarded as an equivalent
tunnelling effect, where the temperature fluctuations enable a reactant
molecule with insufficient energy to overcome the potential barrier.

Ref.2, however, has not taken into account how the fluctuation grows
in the course of time leading to ignition. It is also expected that the
ground temperature rises during this period, which will act coopera-
tively towards reducing the induction time. Taking these two effects
into account, however, involves a mathematical complexity: That is, the

induction time is no longer identified with an eigenvalue obtained as
the inverse growth rate of a linear system. Instead, a nonlinear set
of equations need to be solved in which reaction rates vary as depending
on the temperature elevation as well as on the amplified fluctuation
intensity. These effects will be looked into in what follows.

II. GOVERNING EQUATIONS

We consider a uniform quiescent gas with a temperature elevated suddenly to the level of the ignition temperature. Under this circumstance
which simulates a situation behind a reflected shock wave, an instability, thermal as well as chemical in nature, sets in. In view of experimental observation that numbers of reactive spots are spontaneously and
randomly spawned in the uniform medium, we may treat the stochastic deviation in thermodynamic variables thereby caused as turbulent fluctuations,
which we claim as obeying the following set of equations:

$$\partial_t q_0 + q_{i,i} = 0 \tag{2.1}$$

$$\partial_t q_i + \partial q_{40} / \partial x_i = 0 \tag{2.2}$$

$$\partial_t q_{40} + a^2 q_{i,i} = -(\gamma-1) \Sigma Q_\alpha q_{w\alpha}, \quad (\alpha=1,2,\ldots,N) \tag{2.3}$$

$$\partial_t q_\alpha = q_{w\alpha} \tag{2.4}$$

$$q_{40} = q_4 + \gamma^{-1} a^2 q_0 \tag{2.5}$$

together with the following definitions

$$\partial_t \equiv -i\omega + \partial / \partial t \tag{2.6}$$

$$q_{i,i} \equiv \partial q_i / \partial x_i$$

$$q_{w\alpha} \equiv [(\nu-1)q_0 + (E/RT)\gamma a^{-2} q_4 + \Sigma_{\alpha'} q_{\alpha'} / Y_{\alpha'}] W_\alpha / \rho \tag{2.7}$$

where q's are (smoothed) variables corresponding to (instantaneous)
fluctuation ΔZ's for thermodynamic variables Z's as shown below;

$$q_0 \longleftrightarrow \Delta\rho$$

$$q_1 \longleftrightarrow \rho\Delta u_1$$

$$q_4 \longleftrightarrow \rho R_M \Delta T$$

$$q_{40} \longleftrightarrow \Delta p$$

$$q_\alpha \longleftrightarrow \rho\Delta Y_\alpha \tag{2.8}$$

In the equations above, ρ, u_i, T, p, Y_α and R_M denote the density, the
fluid velocity, the temperature, the pressure, the mass fraction of
species α, and the gas-constant per unit of mass. Also, a, Q_α, W_α,
γ, ν and E stand for the speed of sound, the specific heat of formation

of species α, its mass production reaction rate, the mean adiabatic index, the order of reaction, and the activation energy. Wherever W_α is given by a sum of several terms, namely, $W_\alpha = \sum_{\alpha'} W_{\alpha\alpha'}$, expression (2.7) be replaced by

$$q_{w\alpha} = \sum_{\alpha'} [(\nu_{\alpha\alpha'}-1)q_0 + (E_{\alpha\alpha'}/RT)\gamma a^{-2}q_4$$

$$+ \sum_{\alpha''} (q_{\alpha''}/Y_{\alpha''})]W_{\alpha\alpha'}/\rho$$

Quantities Z's and q's are related through the following expression

$$<\Delta Z_m \Delta \hat{Z}_n> = \int q_m \hat{q}_n^* d\omega \qquad (2.9)$$

which defines a turbulent correlation between thermodynamic quantities Z_m and $\hat{Z}_n = Z_n(\hat{x})$ in terms of an spectral integral of separated variables $(q_\alpha$ and $\hat{q}_\beta)$ for respective points. Integral variable ω is a nondimensional quantity depending only on the frequency, and asterisk(*) denotes the conjugate complex.

The set of Eqs.(2.1) through (2.7) is a linearized version of the original one as derived in ref.3. In the linearized form, and by putting $\omega=0$, the equations are degenerated to linear-perturbed gasdynamic equations with chemical reactions.

These equations are coupled with the ones governing the ground temperature elevation and the species evolution,

$$dT/dt = -\sum Q_\alpha W_\alpha / \rho c_p \qquad (2.10)$$

$$dn_\alpha/dt = W_\alpha \qquad (2.11)$$

also with the reaction rate formula subject to temperature fluctuations (refs. 2,4)

$$W_\alpha(T) = [W_{0\alpha}(T+\delta T) + W_{0\alpha}(T-\delta T)] / 2. \qquad (2.12)$$

In the above expression, $W_{0\alpha}$ denotes the Arrhenius reaction rate and δT stands for the root-mean-square of the temperature fluctuation $<(\Delta T)^2>^{1/2}$ as defined by (2.9). Turbulent reaction rate (2.12) is derived from nonequilibrium statistical mechanics as the exact expression for exchange reaction of the type $A + B \rightarrow C + D$, with a finite activation energy. It applies to all the reactions listed in Table I except for reaction No.6 caused by a triple collision.

In the previous paper[2], the induction time was calculated on the basis of the linear Eqs.(2.11) for radical species which are assumed as varying like $\exp(\lambda t)$. The induction time, which is proportional to λ^{-1}, and is obviously dependent on the ground temperature T_0 through W_α, turns out to be considerably sensitive to the temperature fluctuation δT for $T_0 < 1000$ K. Instead of regarding δT as a fixed parameter as dealt with so in ref.2, we will attempt to solve it in the following.

III. NONLINEAR MECHANISM IN IGNITION OF HYDROGEN-OXYGEN MIXTURES

In the discussion to follow, we employ reactions as listed in Table I which are widely accepted as a minimal sufficient set of elementary reactions to govern the ignition phenomena of hydrogen-oxygen mixtures.

			Q	A	E
0	$H_2 + O_2$	$\rightarrow 2OH$	18.9	1.0×10^{17}	71.4
1	$H_2 + OH$	$\rightarrow H_2) + O$	-15.1	2.19×10^{13}	5.15
2	$O_2 + H$	$\rightarrow OH + O$	16.9	$2.24 \times 10^{14}T$	16.8
3	$H_2 + O$	$\rightarrow OH + H$	1.97	1.82×10^{10}	8.9
6	$O_2 + H + M$	$\rightarrow HO_2 + M$	-47.1	7.95×10^{15}	-1.0
11	$H_2 + HO_2$	$\rightarrow H + H_2O_2$	2.29	1.26×10^{14}	22.0
12	$H_2 + HO_2$	$\rightarrow H_2O + OH$	-53.3	2.0×10^{11}	24.0

Table I. Key reactions in the ignition process used in the present calculation (Q(kcal/mol); heat of formation, A($6:cm^6/mol^2/sec$ others:$cm^3/mol/sec$); frequency factor, E(kcal/mol); activation energy)

Then we need to solve the equations for the mean species concentrations and the mean temperature, respectively,

$$\frac{d}{dt}\begin{bmatrix} n_{OH} \\ n_H \\ n_O \\ n_{HO2} \end{bmatrix} = \begin{bmatrix} K_1 & K_2 & K_3 & K_{12} \\ K_2 & -(K_2+K_6) & K_3 & K_{11} \\ 0 & K_2 & -K_3 & 0 \\ 0 & K_6 & 0 & -(K_{11}+K_{12}) \end{bmatrix} \begin{bmatrix} n_{OH} \\ n_H \\ n_O \\ n_{HO2} \end{bmatrix}, \quad (3.1)$$

$$dT/dt = -\Sigma Q_\alpha W_\alpha / \rho c_p, \quad (3.2)$$

coupled with the set of Eqs.(2.1) through (2.5) which govern fluctuations. hese two groups of equations are coupled nonlinearly through the expression (2.12) for W_α where both the varying temperature and growing temperature fluctuation intervene.

These set of equations constitute an initial-value problem for prescribed initial values of the hydrogen concentration, the ground temperature, the concentration of hydroxyl radical, and the temperature fluctuation which is produced by a shock wave incoherently in the process of
elevating the gas temperature to the level close to ignition.

A calculated result is shown in Fig.1 for the range of interest in
ground temperatures with varied initial values of temperature fluctuations, and for initial hydroxyl radical concentration fixed at $n_{OH}=10^{-14}$.

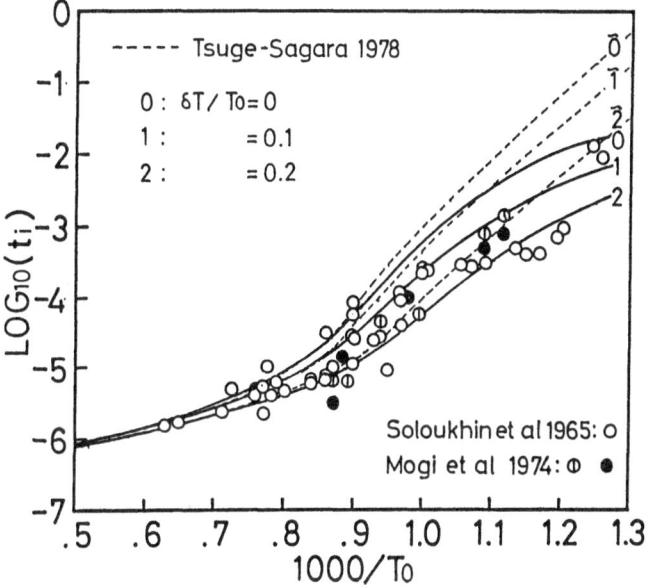

Fig.1 Dependence of Induction Time t_i for
H_2-O_2 Mixtures on Initial Temperature
T_0 (P=2atm)

This value is chosen to be compatible with the existing result by the
linear theory and with the experiment; the linear theory defines the
induction time t_i as $\exp(\lambda t_i) = (n_{OH})_{observable} / (n_{OH})_{initial} = 10^5$,
and that the minimum detectable emission of hydroxyl radical is
$(n_{OH})_{observable} = 10^{-9}$ (mol/cc). Figure.1 shows how the induction time
calculated along this line improves the previous result. This is due
to dual causes: Deviation of the solid curve with index 0 from the
dashed one with the same index reflects the effect due to rise in the
background temperature. The solid curve tends to approach to the experimental values, but is not sufficient to ensure satisfactory agreement
with the data. It is seen that the stochastic temperature fluctuation
which is assumed to exist initially and grows in the course of time
leading to ignition enhances accelerates the chemical reaction to a

greater extent and improves agreement with the experimental data.

Figure.2 depicts how the background temperature and the temperature fluctuation rise between the instant of shock wave passage and the time of ignition. It is seen that major contribution to the term $T + \delta T = T_0 + \Delta T + \delta T$, namely, the most crucial quantity appearing in (2.12) in elevating

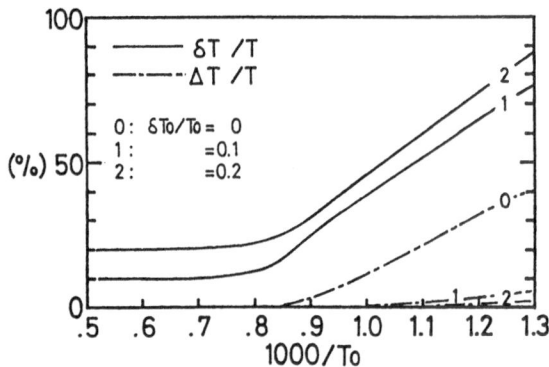

Fig.2 Increase in the background temperature (ΔT) and the temperature fluctuation (δT) during pre-ignition peoriod

the chemical reaction rate is attributed to the amplified temperature fluctuation δT although the rise in the background temperature ΔT plays a considerable role only for cases where no initial temperature fluctuation exists.

IV. CONCLUSIONS

To look into the problem of the classical weak-ignition anomaly more closely, the turbulent reaction rate (formula (2.12)) proposed by Tsugé and Sagara (1978) is used in connection with nonlinear equations of fluctuation evolution. Growth in the temperature fluctuation, if such stochastic one exists initially at all, leads to enormous enhancement of chemical reaction rate and results in further reduction in the induction time compared to the foregoing crude theory in favor of the existing experiments.

References

1. V.V.Voevodsky and R.I.Soloukhin, <u>Tenth Symposium (International) on Combustion.</u> p.279 (The Combustion Institute) 1965.
2. S.Tsugé and K.Sagara, Combustion Science and Technology. 18,179 (1978)
3. S.Tsugé, Physics of Fluids <u>27</u>,1370 (1984)
4. K.Sagara and S.Tsugé, Physics of Fluids <u>25</u>,1970 (1982)

A Lecture on Detonation-Shock Dynamics

D. S. Stewart

University of Illinois, Urbana, Illinois, 61801

J. B. Bdzil

Los Alamos National Laboratory, Los Alamos, New Mexico 87545

Abstract

We summarize some recent developments of J. B. Bdzil and D. S. Stewart's investigation into the theory of multi-dimensional, time-dependent detonation. These advances have led to the development of a theory for describing the propagation of high-order detonation in condensed-phase explosives. The central approximation in the theory is that the detonation shock is weakly curved. Specifically, we assume that the radius of curvature of the detonation shock is large compared to a relevant reaction-zone thickness.

Our main findings are: (1) the flow is quasi-steady and nearly one dimensional along the normal to the detonation shock, and (2) the small deviation of the normal detonation velocity from the Chapman-Jouguet (CJ) value is generally a function of curvature. The exact functional form of the correction depends on the equation of state (EOS) and the form of the energy-release law.

1. Introduction

In this lecture we will describe a theory for unsteady, unsupported, multi-dimensional detonation propagation for the standard explosive model; the reactive Euler equations for a prescribed EOS and rate law. For this model, the detonation structure is ZND, i.e., a shock followed by a reaction zone which contains an embedded, trailing sonic locus. See Figure 1.

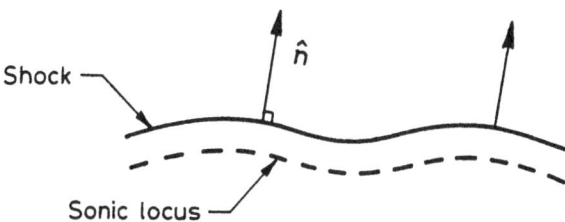

Figure 1. A schematic representation of the detonation shock with normal and trailing sonic locus displayed.

In laboratory frame coordinates, the governing equations for this model are

$$\frac{D\rho}{Dt} + \rho(\nabla \cdot u) = 0 \quad , \tag{1}$$

$$\rho \frac{Du}{Dt} = -\nabla P \quad , \tag{2}$$

$$\frac{DE}{Dt} + P \frac{D(1/\rho)}{Dt} = 0 \quad , \tag{3}$$

$$\frac{D\lambda}{Dt} = r \quad , \tag{4}$$

where in the above ρ, u, P, E, λ and r are respectively the density, particle velocity, pressure, specific internal energy, single reaction progress variable and the rate of forward reaction. To complete the specification of the problem we need to choose constitutive relations for the internal energy function $E(P, \rho, \lambda)$ and the rate law $r(P, \rho, \lambda)$. For illustrative purposes we select the polytropic form for E,

$$E = \frac{P}{\rho}(\gamma - 1)^{-1} - q\lambda \tag{5}$$

where γ is the polytropic exponent, and q is the specific heat of reaction. The solution of these equations must satisfy the standard normal shock relations at the leading detonation shock.

The theoretical developments are carried out in the limit that the radius of curvature of the shock front (R) is much greater than a characteristic reaction-zone length (r_ℓ), i.e.

$$\delta^2 \equiv |r_\ell / R| << 1 \quad . \tag{6}$$

With appropriate assumptions, the main result is that the velocity of the leading detonation shock along its normal deviates from the Chapman-Jouguet value by a small amount that is proportional to curvature (in the simplest cases) and more generally is a function of curvature, i.e.

$$D_n = D_{CJ} - \alpha\kappa \text{ where } \alpha = \text{ constant or } \alpha = \alpha(\kappa) \quad . \tag{7}$$

We were led to the discovery of (7), by our desire to formulate a rigorous theory of the evolution of the detonation shock in complex, two-dimensional (2D) and three-dimensional (3D) geometries, which retained full reaction-zone effects and time dependence, and which was a physically correct and simple-to-use method for correcting detonation velocity. This study was aimed at gaining a fundamental understanding of multi-dimensional detonation.

Our theory is closely related to Whitham's theory of Geometrical Shock Dynamics [1]. Similarly, our theory stresses the dynamics of the shock. However, unlike Whitham, we have a systematic theory of the following flow that supports the shock that is strictly valid when the radius of curvature is large compared to the reaction-zone length.

In Section 2, we give a brief history of earlier developments in 2D detonation theory. We sketch the fundamental approximations and our recent theoretical developments in Section 3. In Section 4, we give some examples of fundamental detonation interactions, while in Section 5, we extend our modeling by examining an energy-release rate that is strongly dependent on state. Finally in Section 6, we comment on the practical implications of the theory for explosive engineering.

2. History of the development

The line of the development of the research presented here can be traced back through the work of Wood and Kirkwood [2] in 1954, Bdzil [3] in 1981, and through the recent collaboration of Bdzil and Stewart from 1984 to the present. See references [4] and [5].

The fact that the detonation propagation speed is dramatically affected by diverging geometry is illustrated by a standard experiment in a rate stick. In that experiment, a cylindrical stick confined by an inert tube is ignited at the bottom by means of a planewave explosive lens and a pad of high pressure booster explosive. A nominally plane, overdriven detonation is thus introduced at the bottom of the stick. As time passes, the detonation shock in the stick becomes curved, because the high-pressure flow expands the tube walls into the relative vacuum surrounding the experiment (i.e., room pressure air). As a result, the plane character of the wave is destroyed. When a steady detonation develops in the stick it has an elliptical-like shape. The final steady 2D-detonation velocity can be measured by simple means and is found to be a function of the radius of the stick and the degree of confinement, i.e., tube wall material and thickness. The steady detonation velocity is reduced from the 1D Chapman-Jouguet value, D_{CJ}, by an amount that becomes greater as the radius of the stick, R_s, is reduced (see Figure 2 for a schematic diagram). At some critical radius, experiments using witness plates show that a steady detonation is not propagated in the stick. Presumably some form of extinction occurs.

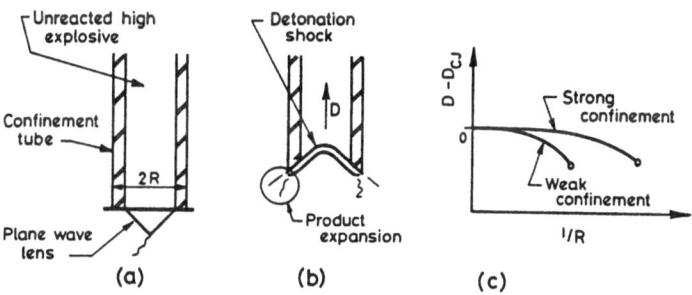

Figure 2. Rate sticks and the diameter effect. Figures 2a and 2b show schematic diagrams of a standard rate stick experiment. Figure 2a shows the stick prior to initiation. Figure 2b shows steady propagation. Figure 2c shows the steady value of the detonation velocity D minus D_{CJ} plotted versus the inverse of the stick radius, R_s^{-1}. Two different cases showing results for strong and weak confinement are shown. The open circles show extinction points which indicate no steady propagation for small radius tubes.

The first theoretical calculations that explained these experimentally observed effects were carried out by Wood and Kirkwood [2]. They used the basic model described in the introduction specialized to a steady, radially symmetric flow. By restricting their analysis to the central stream-line, and by further assuming that the 2D radial flow divergence, $\nabla \cdot u$, was known, they reduced the problem to a system of nonlinear ordinary-differential equations for the steady detonation struc-ture. In particular, they assumed that the quantity, $\nabla \cdot u$ was related to a single *ad hoc* parameter (e.g., R) that measures the divergence of the flow. In these equations the detonation velocity, D, is an unknown constant parameter and R is a specified parameter. Fickett and Davis [6] further showed that this system of equations could be reduced to a single equation for $U^2 \equiv |u - D|^2$, the kinetic energy in the main flow direction, as a function of the reaction progress variable λ.

A qualitative analysis of this governing equation can be carried out quite conveniently in the (U^2, λ)-phase plane. A given value of D defines the starting value for U^2 at the shock. The task is to determine an integral curve in this plane that follows U^2 as λ changes from $\lambda = 0$ at the shock to $\lambda = 1$ at complete reaction. In the limit that the flow divergence is zero, the integral curve terminates at a singular point at $\lambda = 1$. When the flow divergence is non zero, an additional singular point is found in the phase plane that corresponds to the intersection of the thermicity line and the sonic line. The reaction is incomplete at this new saddle-type singular point. The integral curve will pass through this point for only a single value of D for a given R, i.e., $D(R)$. In general, this relationship must be found by numerical shooting techniques. An excellent account of the details of this work is found in Fickett and Davis's book (1979) [6], Section 5g3.

The next contribution to the development of the current theory is due to Bdzil [3]. He analyzed the problem of a steady-state 2D detonation in rate-stick geometry. This analysis was rigorous and not *ad hoc* as was that of Wood and Kirkwood. It was not restricted to the central streamline, but considered the entire 2D problem. This theory is an asymptotic theory which is consistent with the assumption that the stick radius, R_s, is large compared to a 1D reaction-zone length. Once again a parameter equivalent to

$$\delta^2 \equiv |r_\ell / R_s| << 1 \quad ,$$

can be defined. (In Bdzil's account δ is related directly to the angle of the streamline deflection at the confinement boundary.) This assumption is equivalent to a small shock slope, with an $O(1)$ change in the shock position Z_s (measured on the scale of reaction-zone lengths) taking place over the lateral distance scale $r\delta \sim O(1)$ (many reaction-zone lengths).

Bdzil found that all the leading features of the flow could be determined, and that they were parameterized by the shock locus function, Z_s. In turn, the shock locus was a function of the scaled transverse coordinate $\varsigma = r\delta$ and, for a particular example involving the choice of EOS and rate law, satisfied the second-order ordinary-differential equation

$$\frac{D_{CJ}}{2} \left[\frac{dZ_s}{d\varsigma} \right]^2 = \alpha \frac{d^2 Z_s}{d\varsigma^2} - D^{(2)} \quad , \tag{8}$$

where $D^{(2)}$ is identified by the expansion

$$D = D_{CJ} + \delta^2 D^{(2)} \quad ,$$

and measures the deviation of the steady detonation velocity from D_{CJ}.

The position of the shock, Z_s, is measured from a plane, Z = constant, which moves with the steady detonation velocity, D. The function $Z_s(\varsigma)$ determines the local detonation velocity normal to the shock along its extent. Indeed, even though this is not made explicit in Bdzil's paper, equation (8) is equivalent to the coordinate-independent statement

$$D_n = D_{CJ} - \alpha\kappa + o(\kappa) \qquad , \tag{9}$$

where D_n is the velocity along the shock normal. In the above, α is a constant (the assumptions about the EOS and rate law in [3] give α a specific value).

In 1984 we started work on the simplest, most straightforward extension of this steady theory that would include time dependence. We noticed that in order to include time dependence in a quasi-steady theory, it was necessary to introduce a slow-time scale such that the time dependence entered the theory at the same order as the shock curvature. In particular if on the reaction-zone length scale the shock locus, Z_s, is an $O(1)$ function, then the relevant slow-time scale is

$$\tau = \delta^2 t \qquad , \tag{10}$$

where t is measured with the reaction-zone time scale. Calculations with these scaling assumptions show that at leading order, the flow through the reaction zone has the same form as it does in the steady-state problem, i.e., it is quasi-steady. However, the shock locus, which is what parameterizes the solution, is now a function of both the scaled transverse coordinate ς and the scaled time τ.

In contrast to (8), the shock locus, Z_s, now satisfies the partial-differential equation

$$\frac{\partial Z_s}{\partial \tau} - \frac{D_{CJ}}{2}\left[\frac{\partial Z_S}{\partial \varsigma}\right]^2 = \alpha\frac{\partial^2 Z_s}{\partial \varsigma^2} - D^{(2)} \qquad , \tag{11}$$

where Z_s is measured from a constant velocity plane. The above equation is a nonlinear heat equation. Indeed for α = constant, equation (11) can be reduced to a Burgers' equation for the shock slope, $\partial Z_s/\partial \varsigma$. On these length and time scales ς and τ, the evolution of the shock is not governed by a hyperbolic equation, but by the parabolic equation (11). A natural question to ask is why do we find a parabolic evolution equation for a system of hyperbolic equations?

The answer is found in Bdzil and Stewart's [4] (1986) paper on time-dependent 2D detonation. In that paper, we studied the transients that carry an initially 1D detonation into a steady-state 2D detonation. In the example we considered, an initially steady 1D detonation encounters an unconfined corner in the explosive (see Figure (3a)). After the wave reached the corner, the explosive products expanded into the vacuum and the detonation shock began to curve. Because the problem is hyperbolic, a traveling wave head was defined on the detonation shock to the left of which there was no disturbance of the 1D detonation.

We selected the explosive EOS and rate law with the goal of achieving a 1D detonation that was linearly stable to both transverse and flow-direction disturbances. With this goal in mind, we adopted a polytropic EOS model and a rate law for which most of the chemical heat release is given up immediately behind the shock. This was followed by a smaller resolved heat release that took place over a finite distance behind the shock and which controled the dynamics of the problem. For this "small resolved heat-release model," the dynamics of the 1D detonation occur on the "fast" time scale δt. Our results showed that disturbances on the shock propagate according to a hierarchy of two distinct flow regions which occur on the time scales δt and $\delta^2 t$.

In the first region the displacement of the shock is small and the dynamics, which occur on the δt time scale, is wave-like (hyperbolic). This region contains the hydrodynamic wave head, i.e., the leftmost point of the shock disturbance. The magnitude of the shock displacement, length and time scales for this region are given by

$$Z_s \sim O(\delta) \text{ with } \delta^{1/2}r, \delta t.$$

The second region is a diffusion-like region (parabolic). In this region the shock displacement from plane is the largest and the disturbance extends over both the greatest length and time scales. The magnitude of the shock displacement, length and time scales for this region are given by

$$Z_s \sim O(1) \text{ with } \delta r, \delta^2 t.$$

Figures 3a and 3b show a schematic diagram of both the initial configuration and the evolutionary phase of the detonation shock for these two regions.

What we learned from [4] is that the parabolic flow is naturally embedded in the hyperbolic system. The hyperbolic region, while defining the wave head of the disturbance is associated with a small amplitude shock deflection. In contrast the parabolic region is associated with a large scale shock deflection and is the most important region to characterize and measure. The advantage of this description is the relative simplicity of the parabolic region, which involves at most the solution of a simple second-order partial-differential equation (the nonlinear heat equation). Additionally, practical experience with the technologically important case of condensed phase propellants and explosives shows that they have broad well defined detonation shocks. To check the validity of the steady theory for condensed phase explosives, Engelke photographed the shock loci and compared them with the predictions of the steady theory. See Bdzil [3] and Engelke and Bdzil [7]. The theory and experiment were shown to be in qualitative and even quantitative agreement. Therefore, consistency of the unsteady and steady theories then also argues for the parabolic scales.

The results of [4] confirmed the importance of evolution equations of the parabolic type which were discovered earlier. The earlier work was eventually recorded in a paper by Stewart and Bdzil [5], where some examples of relationships between the normal detonation-shock velocity and the curvature were derived for the first time.

The simplicity of the parabolic description makes it possible to do routine calculations of a class of unsteady detonation problems. The detonation-wave spreading problems of greatest interest

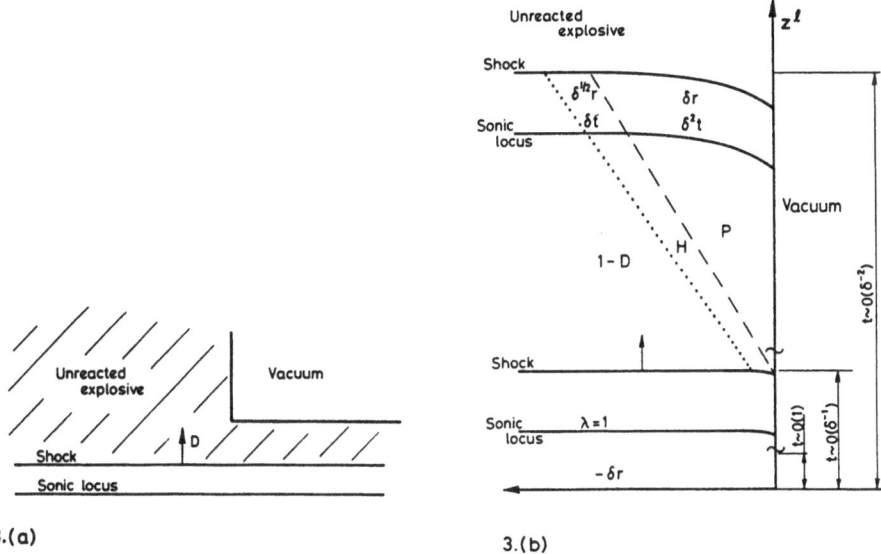

3.(a)

3.(b)

Figure 3. Figure 3a shows the configuration prior to the 1D detonation reaching the vacuum. Figure 3b shows subsequent detonation evolution at two times.

occur in explosives with complicated shapes. If we are to apply the parabolic description outlined above to such problems, we need to carry out the analysis in a system of intrinsic (or problem determined) coordinates. These calculations are the subject of the next section.

3. Sketch of the analysis

In this section we sketch the analysis and explain the approximations used in deriving the shock-evolution equation and the flow description. The model equations are the reactive Euler equations, subject to the shock Hugoniot conditions for a specific EOS and rate law. The presentation here is an outline of the more detailed discussion found in Bdzil and Stewart [8].

The coordinates we choose are shock-attached coordinates, and the problem is three dimensional. Here ξ_i represents arc length along the shock in the directions of the principal curvatures ($i = 1, 2$) defined by the instantaneous shock surface. The variable n represents the distance normal to the shock. The coordinates ξ_i and n form a locally orthogonal coordinate system. A picture of the intrinsic-coordinate system for 2D is shown in Figure 4.

Because we have chosen an intrinsic-coordinate system, the shock curvature appears explicitly

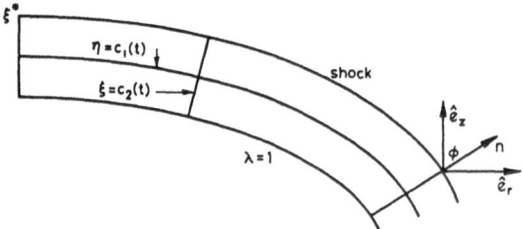

Figure 4. A sketch of the 2D intrinsic shock-attached coordinate system.

in the governing equations of motion. These equations become

Mass:
$$\rho_{,t} - \left[\rho\left(D_n - u_n\right)\right]_{,n} + \kappa\rho u_n + \ldots = 0, \tag{12}$$

Energy:
$$E_{,t} - \left(D_n - u_n\right)E_{,n} - (P/\rho^2)\left[\rho_{,t} - \left(D_n - u_n\right)\rho_{,n}\right] + \ldots = 0, \tag{13}$$

Momentum

$$n: \qquad u_{n,t} + \left(D_n - u_n\right)u_{n,n} + (1/\rho)P_{,n} + \ldots = 0, \tag{14}$$

$$\xi_i: \qquad u_{\xi_i,t} - \left(D_n - u_n\right)u_{\xi_i,n} + \ldots = 0, \quad i = 1,2 \tag{15}$$

Rate:
$$\lambda_{,t} - \left(D_n - u_n\right)\lambda_{,n} = r + \ldots \quad . \tag{16}$$

Note that D_n is the instantaneous shock velocity along the shock normal, u_n and u_{ξ_i} are laboratory-frame particle velocities in the n and ξ_i-directions respectively. The curvature that appears in the above equations is the sum of the principal curvatures, $\kappa \equiv \kappa_1 + \kappa_2$. Higher order terms in the shock curvature are indicated by ellipses.

To these equations we add the shock relations

$$\rho_- D_n = \rho_+\left(D_n - u_{n_+}\right), \qquad P_+ = \rho_- u_{n_+} D_n, \qquad \lambda_+ = 0, \tag{17}$$

$$\frac{D_n^2}{2} = E_+ + \frac{P_+}{\rho_+} + \frac{1}{2}\left(D_n - u_{n_+}\right)^2, \qquad u_{\xi_{i+}} = 0, i = 1,2.$$

The minus subscript refers to the state ahead of the shock, the plus subscript refers to the state behind the shock. In these relations we have adopted the strong shock approximation and have set terms proportional to P_- to zero (we have anticipated that $E_- \sim P_-/\rho_-$).

We make the explicit assumption that the curvature is

$$\kappa \equiv \delta^2 \hat{\kappa}, \qquad \delta^2 \ll 1 \quad , \tag{18}$$

where $\hat{\kappa}$ is the scaled shock curvature and δ^2 measures the magnitude of curvature relative to the 1D reaction-zone length. The length and time scales required are

$$\tau = \delta^2 t, \qquad n, \text{ and } \varsigma_i = \delta\xi_i, \text{ for } i = 1,2 \quad . \tag{19}$$

We introduce the formal expansions for the dependent variables

$$u_n = u_n^{(0)} + \delta^2 u_n^{(2)} + \ldots, \quad u_{\xi_i} = \delta^2 u_{\xi_i}^{(2)} + \ldots,$$
$$P = P^{(0)} + \delta^2 P^{(2)} + \ldots, \quad \rho = \rho^{(0)} + \delta^2 \rho^{(2)} + \ldots, \tag{20}$$
$$\lambda = \lambda^{(0)} + \delta^2 \lambda^{(2)} + \ldots, \quad D_n = D_{CJ} + \delta^2 D_n^{(2)}(\varsigma_i, \tau) + \ldots.$$

Using these expansions in equations (12) – (16) we find that through and including $O(\delta^2)$, the equations that govern the flow reduce exactly to the equations for quasi-steady flow in cylindrical geometry

$$-\left[\rho\left(D_n - u_n\right)\right]_{,n} + \kappa \rho u_n + \ldots = 0, \tag{21}$$

$$\left(D_n - u_n\right) E_{,n} - (P/\rho^2)\left[\left(D_n - u_n\right)\rho_{,n}\right] + \ldots = 0, \tag{22}$$

$$\left(D_n - u_n\right) u_{n,n} + (1/\rho) P_{,n} + \ldots = 0, \tag{23}$$

$$\left(D_n - u_n\right) u_{\xi_i,n} + \ldots = 0, \quad i = 1, 2 \tag{24}$$

$$-\left(D_n - u_n\right)\lambda_{,n} = r + \ldots, \tag{25}$$

since from equation (24) and the shock conditions it follows that $u_{\xi_i} = 0$.

In Section 2 we mentioned that Wood and Kirkwood [2] treated the central streamline problem. Equations (21) – (25) taken together with the normal shock relations are equivalent to the problem they treated. Now, the terms due to the flow divergence are rigorously identified as being proportional to the local shock curvature, κ. The above problem then admits an eigenvalue detonation as its solution. As Wood and Kirkwood showed, it defines a relation between the two parameters D_n and κ as a condition necessary for the integral curve in the (u_n^2, λ)-plane to pass through the saddle singular point, where the flow is sonic. Generally speaking, we have the requirement that there exists a relation of the form

$$D_n = D_n(\kappa) \quad . \tag{26}$$

To illustrate this point we give the equation. Let $U_n \equiv u_n - D_n$, and consider the polytropic EOS

$$E = \frac{P}{\rho}(\gamma - 1)^{-1} - q\lambda \quad . \tag{27}$$

Straightforward manipulation of equations (21) – (25) yields the single ordinary-differential equation for U_n^2 in terms of λ, namely

$$\frac{d(U_n^2)}{d\lambda} = \frac{2U_n^2\left\{q(\gamma - 1)r - c^2\left(D_n + U_n\right)\kappa\right\}}{r\left(c^2 - U_n^2\right)}, \tag{28}$$

where the sound speed is given by $c^2 = \gamma P/\rho = (\gamma - 1)\left[\left(D_n^2 - U_n^2\right)/2 + q\lambda\right]$. The shock boundary condition requires that

$$U_{n+} = -\frac{D_n(\gamma - 1)}{(\gamma + 1)} \quad . \tag{29}$$

Following the nomenclature of Fickett and Davis, the { }-term in the numerator of (28) defines the thermicity locus in the (U_n^2, λ)-plane, and $(c^2 - U_n^2)$ defines the sonic locus. These curves, along with $r = 0$, define the seperatrices and their intersections define the singular points in the phase plane. The object in the phase plane is to find the integral curve that starts from the shock value given by (29) and terminates at complete reaction. Typically such curves must pass through a singular point defined by the intersection of the sonic and thermicity loci. Since κ is small, the intersection point is very close to complete reaction. As mentioned before, this point is a saddle. To ensure passage through the saddle, condition (26) must hold.

In order to give a specific form to relationship (26) we must give the rate law. In Stewart and Bdzil [5] it is shown that for the choice

$$r = k(1 - \lambda)^\nu, \quad \text{for } 0 < \nu < 1, \tag{30}$$

equation (26) takes the form

$$D_n = D_{CJ} - \alpha\kappa + o(\kappa), \alpha \equiv \frac{\gamma^2 D_{CJ}^2}{k(\gamma+1)^2} \int_0^1 \frac{\left(1 + \sqrt{1-\lambda}\right)^2 d\lambda}{(1-\lambda)^\nu} \quad . \tag{31}$$

For the special case of simple depletion ($\nu = 1$) it can be shown that for diverging geometry ($\kappa > 0$)

$$D_n = D_{CJ} + \beta\kappa ln(\kappa) + 2\beta\kappa\left[ln\left(\beta/D_{CJ}\right) - 3\right] + \ldots, \quad \beta \equiv \frac{\gamma^2 D_{CJ}^2}{k(\gamma+1)^2} \quad . \tag{32}$$

4. Detonation interactions

The formulas given in the last part of Section 3 show that the detonation-shock velocity is a function of the curvature of the shock. In order to describe the evolution of the shock we must have a second relation between D_n and κ. Using the surface compatibility conditions of differential geometry, we have derived such a second relation. We call this relation the kinematic-surface condition

$$\frac{1}{\kappa}\left(\frac{1}{\kappa}D_{n,\xi}\right)_{,\xi} + D_n = -\frac{1}{\kappa}\left(\frac{1}{\kappa}\int_{\xi^*}^\xi \kappa_{,t}d\xi\right)_{,\xi} \quad , \tag{33}$$

where ξ^* is a fixed reference position on the shock (see Figure 4). In 2D, the natural representation of the shock locus is in terms of the angle ϕ that the shock normal makes with respect to a fixed reference direction. Then ϕ is related to the shock curvature by

$$\phi \equiv \int_{\xi^*}^\xi \kappa d\xi \quad . \tag{34}$$

If we consider the simple case given by equation (31) and use the scalings given by equation (19), we find that equations (31) and (33) imply the following equation for ϕ,

$$\phi_{,\tau} + \frac{D_{CJ}}{2}\phi\phi_{,s} = \alpha\phi_{,ss} \quad . \tag{35}$$

Equation (35) is Burgers' equation for ϕ. The constant α plays the role of viscosity. Burgers' equation has analytical exact solutions via the Hopf-Cole transformation and its dynamics have been studied extensively. Thus for this example, fundamental shock interaction problems can be studied with these exact solutions. According to our theory, there now exists a catalogue of solutions for detonation-shock interactions, which is similar to the catalogue of solutions to Burgers' equation.

Two simple examples from this catalogue are the step-shock solution and the N-wave solution to Burgers' equation. The step-shock solution corresponds to the solution for two colliding detonations, providing that the detonating material is large enough that the detonation-shock angles are constant in the far field. If two plane detonations are initiated obliquely so as to run into one another, the slope of their common intersected shock locus starts from the left with one value and moves to another value as we pass to the right. Solutions to Burgers' equation show that ultimately a steady-state, step-shock solution is attained with a definite shock-shock [1] thickness that depends on α. This interaction mimics a reactive Mach stem. Importantly, it is diffuse (see Figure 5a).

The N-wave solution corresponds to a positive shock imperfection. In the right and left far field, the detonation is flat and hence ϕ is zero. In the center the shock is raised, giving rise to an N-shape for ϕ, from left to right. The N-wave solution then shows that this imperfection ultimately "diffuses" away; the time required for "diffusion" of the imperfection depends on the value of α (see Figure 5b).

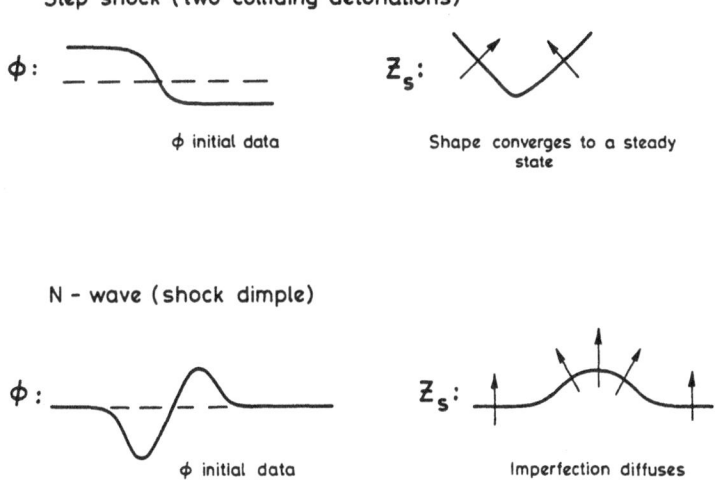

Step shock (two colliding detonations)

ϕ: z_s:

ϕ initial data   Shape converges to a steady state

N - wave (shock dimple)

ϕ: z_s:

ϕ initial data   Imperfection diffuses

Figure 5. Two examples of detonation shock interactions.

5. Stronger state dependence of the rate

The results given by equations (31) and (32) show that the exact functional form of the detonation-shock velocity vs curvature relationship depends on the details of the rate law. Bdzil's [3] results, for steady 2D detonation, showed that as the sensitivity of the rate to the local state is increased, a steady solution does not exist when the curvature becomes sufficiently large. This theoretical observation is consistent with experimental observation.

In this section we present a simple model that shows the consequence of increased state sensitivity. Consider the following shock-state dependent rate (shock-state dependence is typical of solid high explosives)

$$r = kf(\lambda) = \hat{k} \, \exp\left[-\theta\left(D_{CJ} - D_n\right)\right]f(\lambda) \quad . \tag{36}$$

Since D_n is proportional to the shock pressure, the rate multiplier k is now a function of how hard the particles were hit by the passage of the shock. Individual particles react at a rate that is determined by how hard they were shocked. The fact that the state dependence is sensitive (i.e., large changes in r occur for small changes in D_n), is modeled by requiring that the dimensionless parameter

$$\theta D_{CJ} \gg 1 \quad . \tag{37}$$

For the purpose of this illustration, we further consider the following distinguished limit relating the large parameter θD_{CJ} and δ^2

$$\left[\theta D_{CJ}\right]^{-1} = \delta^2 \quad . \tag{38}$$

Using the expansion for D_n, the rate law becomes

$$r = \hat{k} \, \exp\left[D_n^{(2)}/D_{CJ}\right]f(\lambda) \quad . \tag{39}$$

Now it is easy to see that for the case $f(\lambda) = (1-\lambda)^\nu$, where $0 < \nu < 1$, equation (31) still holds, with the exception that k is replaced by $\hat{k} \exp\left[D_n^{(2)}/D_{CJ}\right]$. Using the previous definition for scaled curvature, $\kappa = \delta^2\hat{\kappa}$, we find the reduced shock velocity curvature relation becomes

$$-\left(D_n^{(2)}/D_{CJ}\right)\exp\left[D_n^{(2)}/D_{CJ}\right] = \hat{\alpha}\hat{\kappa} \quad , \tag{40}$$

where $\hat{\alpha}$ is given by equation (31) for α, with $\hat{k}$ replacing k. We rewrite equation (40), in order to compare directly with (31) and (32);

$$D_n = D_{CJ} - \alpha\kappa \, \exp\left[-\theta\left(D_n - D_{CJ}\right)\right] \quad . \tag{41}$$

From equation (41) it is simple to show that for the reduced curvature $\hat{\kappa}$ in the range $0 < \hat{\kappa} < \hat{\kappa}_{cr}$, that there are two values for $D_n^{(2)}$. Hence the detonation velocity is multivalued for positive (divergent) curvature below a critical value of curvature (see Figure 6). For values of curvature above the critical value, it is not possible to have detonation-shock evolution described by the

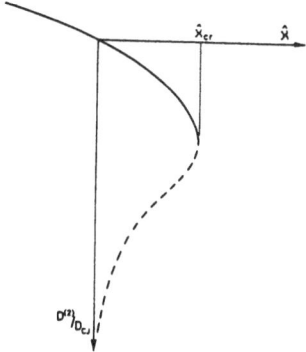

Figure 6. Scaled detonation velocity $D_n^{(2)}/D_{CJ}$ versus scaled detonation shock curvature $\hat{\kappa}$.

parabolic scales. A possible consequence of this is extinction of the detonation wave on portions of the curve where the critical curvature is exceeded.

6. Practical implications for explosive engineering

The theory discussed in this lecture pertains to explosive materials in which a broad, well-defined detonation shock is observed in the limit that the radius of curvature is large compared to the distance from the leading shock to the sonic locus. Indeed this is the case of practical interest for a wide class of explosives.

Engineers who design explosive charges typically use the Huygen's rule of detonation propagation whereby the detonation shock is advanced along its normal at the constant Chapman-Jouguet velocity. Our results indicate that this "recipe" should be modified, and that the correction factor is generally a function of the curvature. In addition our results show that the detonation structure from shock to sonic locus is easily calculated and is locally a 1D, cylindrical, quasi-steady flow.

The theory then suggests that the $D_n(\kappa)$ relation may describe the shock evolution for certain explosives for a wide range of initial and confinement conditions. If this theoretical statement is true, then $D_n(\kappa)$ can be determined directly from experiment. For example, $D_n(\kappa)$ could be determined from photographs of steady detonation-shock loci in rate sticks. Suppose the steady detonation velocity, D, along the axis of the stick has been measured. If ϕ is the angle that the shock normal (taken from the photograph) makes with the axis of propagation, then the normal velocity is given by

$$D_n = D \cos \phi \ .$$

The shock curvature κ could be inferred from the photograph as well. Thus for the extent of the shock locus shown in the photograph, a portion of the $D_n(\kappa)$ curve can be constructed.

Other experiments, steady or unsteady, in totally different geometries, properly analyzed, should reproduce the same $D_n(\kappa)$ in regions of overlap. Consider the case of a 1D, unsteady cylindrically or spherically expanding detonation. In this experiment D_n is simply $\dot{R}$, the rate of change of the radius from the central point, while $\kappa = 1/R$.

Thus the experimentally determined $D_n(\kappa)$ curve, would determine the detonation characteristic for many different geometries and configurations without our having detailed knowledge of either the equation of state or the energy-release law.

References

1. Whitham, G. B. (1974), Linear and Nonlinear Waves, Wiley.

2. Wood, W. W. and Kirkwood, J. G. (1954), Diameter effect in condensed explosives: The relation between velocity and radius of curvature in the detonation wave. J. Chem. Phys. 22: 1920-1928.

3. Bdzil, J. B. (1981), Steady-state two-dimensional detonation, J. Fluid Mech., 108, 195-226.

4. Bdzil, J. B. and Stewart, D. S. (1986), Time-dependent two-dimensional detonation: The interaction of edge rarefactions with finite length reaction zones, J. Fluid Mech., 171, 1-26.

5. Stewart, D. S. and Bdzil, J. B. (1986), The shock dynamics of stable multi-dimensional detonation. University of Illinois, Theoretical and Applied Mechanics Report No. 481, to appear in Combustion and Flame.

6. Fickett, W. and Davis, W. C. (1979), Detonation, University of California Press, Berkeley.

7. Engelke, R. and Bdzil, J. B. (1983), A study of the steady-state reaction-zone structure of a homogeneous and heterogeneous explosive. Phys. Fluids, 26(5), 1210-1221.

8. Bdzil, J. B. and Stewart, D. S. (1987), Detonation Shock Dynamics, submitted for publication.

Acknowledgments

Earlier versions of this lecture have been presented by J. B. Bdzil and D. S. Stewart: (1) International Workshop on Mathematics in Combustion (IWOMIC)|Garmish-Partenkirchen, FDR (August 1986) and (2) SIAM Conference on Numerical Combustion|San Francisco, USA (March 1987). An earlier account appears as Cranfield report, CoA Report No. NFP/8707, August, 1987. D. S. Stewart gratefully acknowledges the hospitality of Professor John Clarke and the Cranfield Institute of Technology, and support from Contract No. NNS/32A/1A91965 (AWRE, Aldermaston). D. S. Stewart is supported by a contract with Los Alamos National Laboratory (DOE-LANL-9xr6-5128c1). J. B. Bdzil is supported by the U.S. Department of Energy (DOE-W-7405-ENG-36).

2.   Complex Chemistry and Physics

ASYMPTOTIC ANALYSIS OF BRANCHED-CHAIN IGNITION
IN THE COUNTERFLOW FIELD

Takashi Niioka

National Aerospace Laboratory, Kakuda Branch

P.O. Box 7, Ohgawara, Miyagi 989-12, JAPAN

Abstract

Asymptotic analysis of ignition in a typical stretched flow field counterflow system is presented.   First,   previous theoretical treatment of overall chemical reaction kinetics and related experimental results are summarized.   Second, analysis of the branched-chain reaction mode of ignition in the same flow field is presented.   The results show that the ignition time increases at a strong stretch rate and high recombination reaction rate.

1. INTRODUCTION

The most fundamental and typical stretched-flow field is provided by the counterflow system.   This system is relatively convenient for observation of the interaction between a chemical reaction and flow mechanics in the laboratory scale and also for treatment of the analysis as a one-dimensional phenomenon.   It has, therefore, been extensively used for extinction analyses and experiments dealing with flame established in such flow fields during the past two decades, and the effect of the Damköhler number (inversely proportional to the stretch rate) has been revealed through these studies.   Although ignition has also been investigated, the analyses have dealt with steady treatment as well as extinction.   Therefore, only the conditions necessary for ignition time have been obtained, but the effect of stretch rate on ignition time has never been analyzed.   Experiments on ignition in this flow configuration have also been very rare and no ignition data except the authors' results [1-3] have existed for the low Damköhler number region in spite of the fundamental and practical importance of convective ignition.

The most prominent characteristic of ignition time in this flow configuration is that it increases as the stretch rate increases. To show this feature theoretically, the author carried out asymptotic analysis[4]. Based on this analysis, anomalous experimental ignition times were elucidated and compared with theoretical results based on this analysis. Up to now, the chemical reaction used to analyze these ignition characteristics has been the overall one-step reaction. Here, however, we use a two-step chemical reaction to analyze ignition in the same counterflow field.

2. PREVIOUS ANALYSES AND EXPERIMENTS

The schematic explanation of the problem solved is shown in Fig. 1. Two opposed-jet inviscid flows with equal velocity impinge on an adiabatic thin plate and the plate is removed at time zero. One-step irreversible gas-phase reaction ($F + n\bar{O} \rightarrow$ Products) is assumed, and the Prandtl number and the Lewis number are taken as unity for simplicity. Furthermore, if we assume the stream function (f) to be equal to the nondimensional distance (η), the energy and mass conservation equations are simplfied as follows [4]:

$$\mathscr{L}\{Y_F\} = \mathscr{L}\{Y_{\bar{O}}\} = -\mathscr{L}\{\theta\} = -\Lambda_0 Y_F Y_{\bar{O}} exp(-E/\theta), \qquad (1)(2)(3)$$

Fig.1 Schematic showing the flow configuration under consideration.

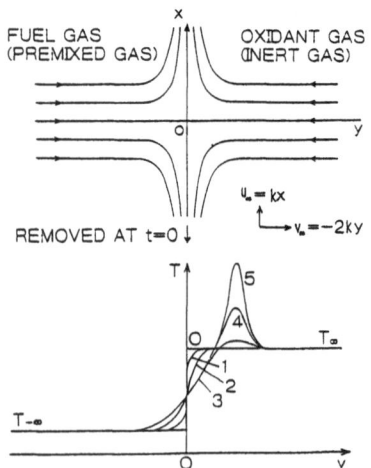

where the operator is $\mathcal{L} \equiv \dfrac{\partial}{\partial \tau} - \dfrac{\partial^2}{\partial \eta^2} - \eta \dfrac{\partial}{\partial \eta}$. $\qquad$ (4)

The symbols are listed at the end of this paper and are different from those used in the author's previous paper[4] in order to coincide with the following analysis for the two-step reaction scheme.

Gas-phase ignition time is given as

$$\tau_{ig} = -\frac{1}{2} \ln \left[1 - \frac{4\theta'(2-\theta')\ln(E\theta'/2\sqrt{\pi}\theta_\infty^2)}{\Lambda_0 e^2 (1-\theta')^2 exp(-E/\theta_\infty)} \right] ,$$ $\qquad$ (5)

where $\theta' = \theta_\infty - \theta_{-\infty}$, $\qquad$ (6)

for the case of initially unmixed reactants. When the counterflow velocity increases, the Damköhler number (Λ_0) decreases, and therefore ignition time increases. This is because the exothermic reaction slows down owing to the stretched-flow and the subsequent temperature rise does not proceed rapidly. At a certain value of Λ_0 or k, the ignition time extends to infinity and the critical Damköhler number is obtained by substituting zero for the portion in brackets.

The experiments[1-3] were conducted by using condensed fuels in a hot stagnation-point flow because the two opposed gas-jet system is quite difficult to control in the ignition experiment. In the lower velocity range, ignition times are nearly equal to gasification time and therefore become shorter as hot oxidant flow velocity increases. At higher velocities, however, they should lengthen as the velocity increases, since according to the gas-phase ignition theory, ignition in the stretched-flow can not occur rapidly even if gasification is completed. A comparison of the measured ignition time with the calcu-

Fig.2 Comparison of theoretical ignition time with measured results.

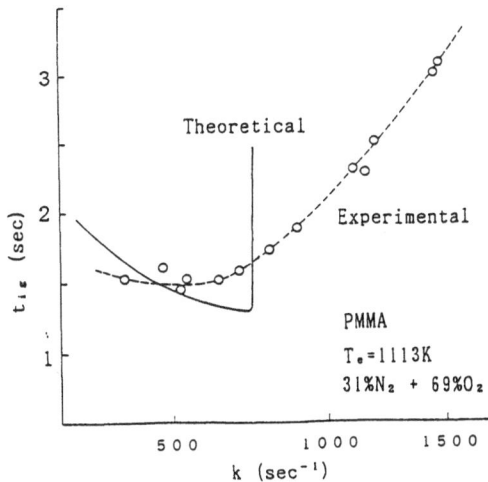

lated results obtained from Eq.(5) is shown in Fig. 2. The above-mentioned feature of the minimum ignition time is verified. Measured points are the mean values of several tests, but the infinite ignition time whose number is apt to increase in a region of higher velocity was not included in the calculation of the mean values. This is the reason why the agreement is weaker at the higher stretch rate.

Using the above-mentioned characteristic ignition time behavior, the interesting features of the ignitable limit of mixed fuel[2,3] and of the complex fuel spray ignition [5] could be precisely explained.

The case of hot inert gas (for example, burned gases) versus cold premixture (see Fig.1) was also analyzed[4]. The nondimensional ignition time in terms of the nondimensional stretch rate is shown in Fig. 3. The basic ignition time tendency is the same as that in the case

Fig.3 Theoretical prediction
of ignition time for overall
chemical reaction of premixture as a function of nondimensional stretch rate.

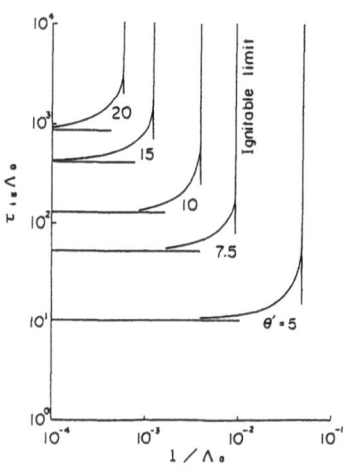

of initially unmixed gases: Ignition time lengthens as the stretch rate increases. This premixture case is extended to the case of a two-step chemical reaction in the next chapter.

3. IGNITION UNDER TWO-STEP REACTION KINETICS

3.1 Simplified Problem

We consider ignition of a premixture supported by a two-step chemical reaction in the stretched-flow field. Ignition or flame initiation of a cold combustible gas stream coming in contact with a hot inert gas stream was presented in the author's previous paper[6].

The system under consideration here is illustrated in Fig. 1. The idealized two-step reaction[7,8] is adopted in this analysis: a reactant (a) and an intermediate radical species (b) generate two radicals through a chain-branching reaction (reaction I: a+b → 2b), and then the radicals recombine into final combustion products (p) through a chain-termination reaction (reaction II: b+b+M → p+M). This idealized kinetic scheme may be applicable to the hydrogen-oxygen reaction. Reactant (a) indicates a deficient species of the premixture, and the representative radical (b) depends on whether the mixture is rich or lean, or whether the phenomenon is flame propagation or ignition.

In reaction I, the temperature dependency is large and therefore the Arrhenius reaction term must be taken into account, although the reaction heat can be neglected. On the other hand, the activation energy can be assumed to be zero in reaction II since the temperature dependency is quite small, but the recombination reaction heat must be considered. Then, under the same assumptions as described in the derivation of Eqs. (1-3), the governing equations can be set as follows:

$$\mathscr{L}\{\alpha\} = -\Lambda_1 \alpha\beta exp(-E/\theta), \tag{7}$$

$$\mathscr{L}\{\beta\} = \Lambda_1 \alpha\beta exp(-E/\theta) - \Lambda_2 \beta^2, \tag{8}$$

$$\mathscr{L}\{\theta\} = \Lambda_2 \beta^2, \tag{9}$$

where the operator is also the same as that of Eq.(4). The boundary and initial conditions are

$$\alpha(0, \eta \geq 0) = \alpha(\tau, \infty) = 0, \tag{10}$$

$$\alpha(0, \eta < 0) = \alpha(\tau, -\infty) = 1, \tag{11}$$

$$\beta(0, \eta) = \beta(\tau, \pm\infty) = 0, \tag{12}$$

$$\theta(0, \eta \geq 0) = \theta(\tau, \infty) = \theta_\infty, \tag{13}$$

$$\theta(0, \eta < 0) = \theta(\tau, -\infty) = \theta_{-\infty} . \tag{14}$$

The stretch rate (velocity gradient in the direction of the opposed-jet axis) is included in the two parameters (Damköhler numbers). Therefore, it is necessary to examine the effect of Damköhler numbers on the ignition time.

3.2 Activation Energy Asymptotics

The analysis of the system parallels that of the previous paper for the co-flowing system[6]. The case of $\Lambda_1 exp(-E/\theta_\infty) \gg \Lambda_2 \gg \Lambda_1 exp (-E/\theta_{-\infty})$ is most interesting and realistic; the recombination reaction has an intermediate rate between the fast chain-branching reaction rate at θ_∞ and the slow rate at $\theta_{-\infty}$. In this case, a temperature equal to θ^* exists at which the chain-branching reaction and recombination reaction have the same rate, and the next relation holds at $\eta = \eta^*$,

$$\Lambda_2 = \Lambda_1 exp(-E/\theta^*). \tag{15}$$

Since reaction I has a strong temperature dependency, the region of $\eta < \eta^*$ must be in a frozen state and that of $\eta \geq \eta^*$ must be in equilibrium ($\alpha=0$). Therefore, the distribution of α for $\eta \leq \eta^*$ can be obtained by setting the right-hand side of Eq.(7) equal to zero:

$$\alpha = erfc \left\{ \frac{g(\eta)}{\sqrt{2(1-e^{2\tau})}} \right\} - 1, \tag{16}$$

where

$$g(\eta) = (\eta-\eta^*) - \frac{\eta^*}{2}(\eta-\eta^*)^2 + \frac{\eta^*}{6}(\eta-\eta^*)^3 \text{ ---------} . \tag{17}$$

The concentration gradient at η^* is equivalent to

$$\frac{\partial\alpha}{\partial\eta} \bigg|_{\eta^*} = - \sqrt{\frac{2}{\pi(1-e^{-2\tau})}} . \tag{18}$$

On the other hand, since the summation of Eqs. (7-9) yields the simple relation

$$\alpha + \beta + \theta = (1 + \theta_{-\infty} - \theta_\infty)\xi + \theta_\infty , \tag{19}$$

the difference of the concentration gradient of β at both sides of η^* can be derived as

$$\frac{\partial\beta}{\partial\eta} \bigg|_{\eta^*-} - \frac{\partial\beta}{\partial\eta} \bigg|_{\eta^*+} = \sqrt{\frac{2}{\pi(1-e^{-2\tau})}} , \tag{20}$$

by using the continuity of temperature at η^*.

Reaction I proceeds only in the vicinity of η^*, and therefore the equation with respect to β,

$$\frac{\partial\beta}{\partial\tau} = \frac{\partial^2\beta}{\partial\eta^2} - \eta\frac{\partial\beta}{\partial\eta} - \Lambda_2\beta^2 , \tag{21}$$

$$\beta(0,\eta) = \beta(\tau, \infty) = 0 \quad \text{for} \quad \eta > \eta^* , \tag{22}$$

$$\beta(0,\eta) = \beta(\tau, -\infty) = 0 \quad \text{for} \quad \eta < \eta^* , \tag{23}$$

is available for both sides apart from η^*. The next step is to solve Eq.(21) and to substitute the result into Eq.(20). Except for θ^* close to θ_∞ the rate of reaction I does not increase because of the large temperature dependency, i.e., the value of η^* must be large enough to enable the value of ξ^* to become very small. Using the stretched variable which makes the distance variable of order unity

$$\zeta \equiv (E/\theta_\infty^2)\xi , \tag{24}$$

Eq.(21) can be reduced to

$$\frac{\partial\beta}{\partial\tau} = \frac{2\zeta^2}{1-e^{-2\tau}} \ell n\left(\frac{E}{2\sqrt\pi\theta_\infty^2}\right) \frac{\partial^2\beta}{\partial\zeta^2} - \Lambda_2\beta^2 , \tag{25}$$

in a first approximation. The derivation of this equation is similar to that presented in the previous paper[4,6].

Compared with the diffusion term, the time derivative term can be neglected in a relative error of order $1/\ell n(E/2\sqrt\pi\theta_\infty^2)$. Thus the region around the reaction zone can be regarded as being in a quasi-steady state. Equation (25) reduces to the ordinary differential equation

$$\zeta^2 \frac{d^2\phi}{d\zeta^2} = \phi^2 , \qquad (26)$$

by use of the new dependent variable.

$$\phi \equiv \Lambda_2 (1 - e^{-2\tau}) \beta / 2 \ln(E/2\sqrt{\pi}\theta_\infty^2). \qquad (27)$$

This equation is exactly the same as that used in the previous paper [6], and the concentration gradient can be derived as follows:

$$\zeta * \frac{d\phi}{d\zeta} \bigg|_{\zeta *_-} - \zeta * \frac{d\phi}{d\zeta} \bigg|_{\zeta *_+} \cong 2\sqrt{\frac{2}{3}} \phi^{*3/2}. \qquad (28)$$

Converting the left-hand side into the derivatives with respect to η and substituting it into Eq.(20), we can obtain an equation including $X \equiv \eta/\sqrt{2(1-e^{-2\tau})}$ and β^*. Since the relation between X and β^* is also derived from Eq.(19) by putting $\alpha = 0$ at η^*, we can solve the problem with these two equations:

$$e^{-2\tau} = 1 - \frac{3\ln(E/2\sqrt{\pi}\theta_\infty^2)}{4\Lambda_2 \beta^{*3}} \{erfc(X)exp(X^2)\}^2, \qquad (29)$$

$$\beta^* = \frac{1}{2} (1 + \theta_{-\infty} - \theta_\infty)erfc(X) + \theta_\infty - \theta^*. \qquad (30)$$

3.3 Calculated Results

We can calculate the variation of β^* with τ by substituting X arbitrarily into Eqs.(29) and (30). As seen in Fig.4, two values of β^* exist for times lower than a maximum time at which the time derivative of becomes infinite, showing a behavior such as thermal runaway. This moment can be considered as being the time of ignition. The

Fig.4 The variation of the radical concentration at η^* with nondimensional time.

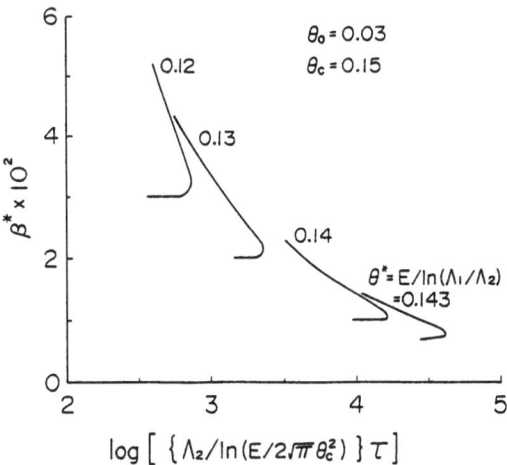

$\theta_0 = 0.03$
$\theta_c = 0.15$
0.12
0.13
0.14
$\theta^* = E/\ln(\Lambda_1/\Lambda_2)$
$= 0.143$

$\beta^* \times 10^2$

$\log \left[\{\Lambda_2/\ln(E/2\sqrt{\pi}\theta_c^2)\} \tau \right]$

Fig.5 Theoretical prediction
of ignition time as a function of reaction rate ratio
for values of the non-
dimensional stretch rate.

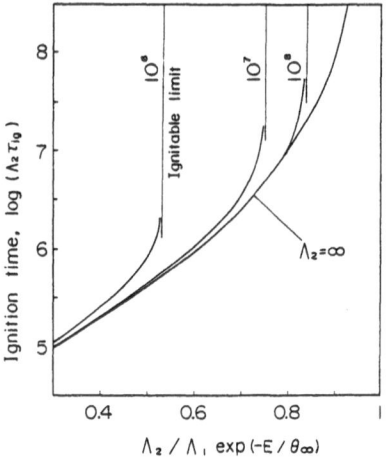

Fig.6 Theoretical pre-
diction of ignition time
as a function of non-
dimensional stretch rate
for values of the reac-
tion rate ratio.

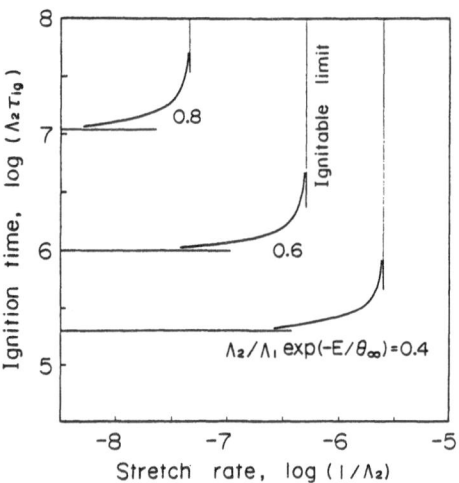

ignition time can also be found in an alternative way: Since we can
determine the approximate value of X_{ig} at ignition by equation

$$X_{ig}^2 \ erfc(X_{ig}) = \frac{2(\theta_\infty - \theta^*)}{3(1 + \theta_{-\infty} - \theta_\infty)} ,$$ (31)

substitution of X_{ig} into Eq.(29) yields the ignition time (τ_{ig}).
Values used for calculation are E=3, θ_∞=0.15, and $\theta_{-\infty}$=0.03.

Figure 5 shows the nondimensional ignition time in terms of the
ratio of the reaction rates, with the ignition time increment being
due to the high rate of the recombination reaction.   The reaction rate
ratio cancels the effect of the stretch rate.   The parameter (Λ_2),
however, is inversely proportional to the stretch rate.

Such as Fig.5, Fig.6 shows ignition time in terms of the non-dimensional stretch rate as the lateral axis, and corresponds to Fig.3 with regards to the overall reaction of fuel and oxygen. The characteristic variation is quite the same although the parameter is different. Large increase of the stretch rate and the recombination reaction rate results in an ignitable limit, and above this point ignition does not take place.

The critical Damköhler number below which ignition is unobtainable is shown in Fig.7, and the value of $\theta*$ and that of the radical concentration at ignition is plotted in Fig.8.

Fig.7 Theoretical prediction of critical Damköhler number.

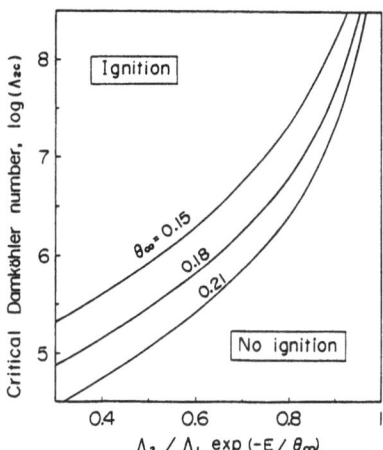

Fig.8 Temperature and radical concentration at $\eta*$ at the time of ignition.

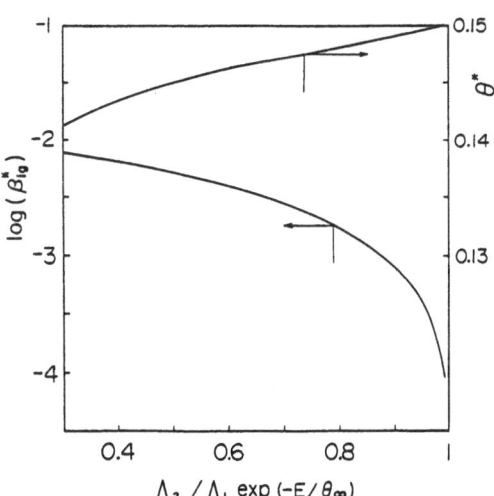

4. CONCLUDING REMARKS

Ignition behavior in the stretched-flow field counterflow system was discussed, especially for the case of branched-chain kinetics. By use of the present analysis, the effect of the stretch rate and the reaction rate ratio of a chain-branching reaction to the recombination reaction was revealed: The increase of the stretch rate and the recombination reaction rate caused the long ignition time.

The present ignition mode must be a fundamental feature of turbulent combustion, especially of the distributed reaction zone, because ignition of very small scale eddies or pockets of premixture surrounded by hot burned gas flowing with strong stretch may play an important role in this tubulent combustion region, as discussed at the US/Japan Joint Seminar herein reported. The present analysis of ignition is expected to be useful for basic treatment when we consider turbulent combustion.

Also, the present theory must be extended to the following challenging problems: more practical ignition problems of solid propellants with a two-step reaction scheme, for example HMX and Double-Base propellants, ignition problems of unmixed systems under two-step reaction kinetics, and the more complicated case of multi-step reactions such as ignition of hydrocarbons.

Nomenclature

a	reactant		$\tilde{Y}$	mass fraction
b	radical		Y_F	$\tilde{Y}_F/\tilde{Y}_{F-\infty}$
B	frequency factor		$Y_{\bar{o}}$	$M_F\tilde{Y}_{\bar{o}}/nM_{\bar{o}}\tilde{Y}_{F-\infty}$
c	specific heat at constant		α	$\tilde{Y}_a/\tilde{Y}_{a-\infty}$
	pressure		β	$\tilde{Y}_b/\tilde{Y}_{a-\infty}$
$\tilde{E}$	activation energy		ζ	$(E/\theta_\infty^2)\xi$
E	$c\tilde{E}/RQY_{F-\infty}$ or $c\tilde{E}/RQY_{a-\infty}$		η	$\sqrt{2k\rho_\infty\lambda_\infty/c_\infty}\int_o^y(\rho/\rho_\infty)dy$
f	$-\rho v/\sqrt{2k\rho_\infty\lambda_\infty/c_\infty}$		θ	$cT/Q\tilde{Y}_{F-\infty}$ or $cT/Q\tilde{Y}_{a-\infty}$
k	stretch rate (velocity gràdient)		θ'	$\theta_\infty-\theta_{-\infty}$
M	molecular weight		λ	thermal conductivity
n	stoichiometric coefficient		Λ_0	$B_0\rho\tilde{Y}_{F-\infty}/2k$
Q	heat of reaction		Λ_1	$B_1\rho\tilde{Y}_{a-\infty}/2k$
R	universal gas constant		Λ_2	$B_2\rho\tilde{Y}_{a-\infty}/2k$

t	time	ξ	$erfc(X)/2$
T	temperature	ρ	density
u.v	velocity (see Fig.1)	τ	$2kt$
x,y	distance (see Fig.1)	ϕ	$\Lambda_2(1-e^{-2\tau})\beta/2\ln(E/2\sqrt{\pi}\theta_\infty^2)$
X	$\eta/\sqrt{2(1-e^{-2\tau})}$		

(suffixes)

a	reactant	1	reaction
b	radical	2	reaction
c	ignitable limit	∞	$y=\infty$
F	fuel	$-\infty$	$y=-\infty$
ig	ignition	*	reaction zone of reaction
$\bar{O}$	oxygen		(see Eq.(15))
0	overall reaction		

REFERENCES

[1] Niioka, T., Takahashi, M. and Izumikawa, M.; Eighteenth Symposium (International) on Combustion, The Combustion Institute, 1981, p741.

[2] Niioka, T., Mitani, T. and Sato, J.; Twentieth Symposium (International) on Combustion, The Combustion Institute, 1985, p1877.

[3] Niioka, T., Mitani, T. and Sato, J.; Transactions of Japan Society of Mechanical Engineers, vol.51, No.467(B), 1985, p2457. (in Japanese)

[4] Niioka, T.; Eighteenth Symposium (International) on Combustion, The Combustion Institute, 1981, p1807.

[5] Sato, J., Konishi, K., Okada, H. and Niioka, T.; Twentyfirst Symposium (International) on Combustion, The Combustion Institute, 1987 (to be published).

[6] Niioka, T.; Combustion and Flame (submitted).

[7] Linán, A.; A theoretical analysis of premixed flame propagation with an isothermal chain reaction, Technical Report No.1, Instituto Nacional De Tecnica Aerospacial "Esteban Terradas", Madrid, Spain, 1971.

[8] Zeldovich,. Y.B.; K Teorii Rasprostramemia Plameni, Zhurinal Fizicheskoi Khimii (USSR), T. 22:27 (1948), Published also as NACA Tech. Memo. 1282, 1951.

ASYMPTOTIC METHODS FOR FLAMES WITH DETAILED CHEMISTRY

F.A. Williams
Department of Mechanical and Aerospace Engineering
Princeton University
Princeton, NJ 08544

I. Introduction

Recently there have been rapid advances in applications of asymptotic methods to describing structures of premixed laminar flames and laminar diffusion flames[1] In the past, asymptotic treatments have been reserved mainly for one-step, Arrhenius approximations to the true chemical kinetics, resulting the well-known activation-energy asymptotics (AEA). These past studies have helped to clarify general aspects of flame structure, premixed-flame propagation velocities, flame extinction, stability, and response to turbulence. There has always been uncertainty concerning the extent to which the predictions may be negated by the complex chemical kinetics of real flames. Overall reaction rates in real flames generally do increase rapidly with increasing temperature, lending confidence to qualitative predictions of burning velocities, extinction, stability, and response by AEA. However, detailed flame structures are known to depend strongly on chemical-kinetic interplay of elementary reaction steps. Hence, AEA may offer poor answers to specific questions about flame structure. Many maintain that AEA is entirely inappropriate even for describing phenomena such as extinction; extinction may result mainly from a change in the chemical-kinetic mechanism[2]. Extensions of asymptotic methods to include aspects of detailed chemistry can help to resolve these uncertainties.

Our knowledge of rates of elementary steps in ideal-gas reactions has continually improved over the years[3]. This improvement has reduced uncertainties in elementary rates to a point at which computations of flame structures, burning velocities, and extinction conditions can be made with reasonable accuracy for comparison with experimental results. Advances in numerical methods for flame computations, along with improvements in computer capabilities, have resulted in entirely reliable flame-structure predictions that agree well with experiment, for some flames; examples are hydrogen flames, lean or stoichiometric methane flames, and lean or stoichiometric flames of mixtures of hydrogen and carbon monoxide, all burning with oxygen, typically in air. In principle it is possible to compare results of these computations with those of AEA, to test the latter. This has not been done, and in practice it is difficult, because the bases of the two approaches differ so greatly. To what kinds of overall steps should AEA be considered to apply? Extensive investigations are needed to address such questions.

The requisite investigations have been initiated recently, not through numerical methods, but rather through asymptotic methods with detailed chemistry. These asymptotic methods consider not only activation energies but also ratios of rates of elementary steps. Thus, in addition to AEA, what might be called methods of rate-ratio asymptotics (RRA) are involved. Applications of mixtures of AEA and RRA have now been made to the ozone decomposition flame, to stoichiometric methane-air flames, and to the methane-air diffusion flame, and some ideas about hydrogen-halogen flames have been developed. Each of these studies has proven to be challenging in its own right and to involve new concepts. Wouldn't it be easier to extract the same information from outputs of numerical integrations? Probably not, because without the ideas that go into the new RRA and AEA, we would not know what questions to ask of the computational results. There seems to be no short-cut to improved understanding of detailed flame structure and of its relationship to one-step AEA predictions. Aspects of flame structures with detailed chemistry, as uncovered by recent asymptotic analyses, are reviewed here.

2. THE OZONE DECOMPOSITION FLAMES[4-7]

The ozone decomposition flame $(2O_3 \rightarrow 3O_2)$ is the simplest of all real flames because it involves only one element and three molecules, O, O_2 and O_3. The reactions steps are $O_3 + M \rightleftarrows O_2 + O + M$, $O + O_3 \rightarrow 2O_2$ and $O + O + M \rightarrow O_2 + M$; the oxygen molecule is so stable that the reverses of the last two generally can be neglected. The flame structure depends on the pressure (P) the initial temperature (IT) and the initial ozone mole fraction (IOMF) of the ozone-oxygen mixture. The main aspects of the structure can be considered to depend mainly on IOMF.

At very low IOMF, or at high enough IT, the combustion occurs in an essentially time-dependent manner, and a laminar burning velocity does not exist, not even with formulations account for reactions ahead of the flame; at ordinary temperatures this regime is well beyond flammability limits by heat loss. There may be a regime, at slightly large IOMF, in which burning velocities can be defined by allowing for reactions ahead of the flame, or by replacing AEA of the first step by a heat-release asymptotics[8] (HRA) in which the Zel'dovich number is not large but an activation temperature based on the initial temperature is; these questions have not yet been fully investigated. At low IOMF in the typical flammable range, a steady state applies for O, and an effective one-step reaction is readily derived[4,5], for which application of AEA produces excellent agreement with results of full numerical integrations. At higher IOMF this steady-state regime is replaced by a merged regime, in which the first two steps occur in a reactive-diffusive zone following the preheat zone, but the steady state is not attained[4-6].

In both of these regimes, at least for $P \leq 50$ atm, the third step and/or the reverse of the first occur only in a downstream zone that maintains a convective-reactive balance and that does not affect the burning velocity. This recombination zone, identified first for this flame, appears to be a potential part of many other flames as well. The end of the flame, insofar as its propagation is concerned, then occurs for kinetic reasons and is not identifiable with the adiabatic flame temperature. This kinetic intrusion into burning velocities is not part of one-step AEA but is entirely consistent with AEA of the ozone mechanism, for example. To the extent that long recombination zones exist, in investigations of nonplanar, time-dependent flames (e.g., in turbulence), it may be necessary to retain distributed reactions in the burnt gas (but not in the unburnt gas) for accuracy in numerical integrations.

In the merged regime, the burning velocity from a two-term AEA expansion is appreciably below that of the one-term expansion, and the results from the full number integration fall between. A one-third rule has been proposed[7,9], in which the burning velocity is estimated by returning one third of the way from the two-term value to the one-term value. This produces both excellent agreement with results of numerical integrations and strong criticisms by applied mathematicians. Clearly it is in no way rigorous; the AEA expansions have not even been proven oscillatory. Yet, experience with other flames (e.g., certain one-step AEA flames suggests that in fact oscillatory approximation is a general attribute of AEA expansions, and the one-third rule is a useful practical tool if we are forced to guess a burning velocity with only two-term AEA results available. It appears that a problem worthy of mathematical study is the identification of sufficiency conditions for oscillatory behavior of AEA approximations.

Possibly HRA could remove the poor convergence of AEA in the merged regime. If so, the difficulty should be shrinkage of the preheat zone, since the HRA flame is essentially one with a convective-reactive-diffusive balance nearly throughout (and therefore one not readily facilitating response analyses). It is uncertain whether this is the difficulty or whether accounting for reactions ahead of the flame (unsteadiness) is essential; more research could be helpful.

3. PREMIXED HYDROGEN-HALOGEN FLAMES

The overall chemistry of hydrogen-halogen flames is $H_2 + Ha_2 \rightarrow 2HHa$, where Ha = F, Cl, Br, I. Usually attention is focused on Br or Cl as Ha, and there is a well-known chain mechanism, with H and Ha as carriers. These flames are not yet well explored by asymptotics, but with the exception of Ha = I, which may never involve a carrier, it seems clear that regimes will exist in which full steady states for intermediates apply, and AEA based on $Ha_2 + M \rightarrow 2Ha + M$ will work. The extent of analogy with the ozone flame is surprising[1]. Studies of ozone showed a possible two-zone structure, in which the O_3 dissociation occurred in a hot downstream zone and the second, exothermic, step occurred in which was the preheat zone through upstream diffusion of O, both reaction zones now being convective-reactive-diffusive[6]. But, according to the numbers, this structure never occurs for the ozone flame. However, for Ha = F, or for Ha = Cl, the energy carried by Ha may be large enough, and the activation energy for $Ha + H_2 \rightarrow H + HHa$ may be small enough, that there may be conditions under which the two-zone structure exists. The two-zone structure gives an asymptotic description of flame propagation by the long-debated mechanism of upstream diffusion of active reaction intermediaries and shows that this mechanism is consistent with a suitable multi-step AEA. Hydrogen-halogen flames are in need of much more study by asymptotic methods to ascertain the occurrence of this mechanism and to find whether other mechanisms may arise.

4. PREMIXED HYDROCARBON FLAMES

Compared with the flames discussed thus far, hydrocarbon flames pose stiff challenges for asymptotic methods. The simplest hydrocarbon flame is the methane-air flame, and numerical methods for describing this flame have employed more than 200 elementary reaction steps. It seems impossible to approach an asymptotic analysis of detailed-chemistry structures of hydrocarbon flames without any preconceived ideas of the flame structure in mind. Aside from background prejudices, a good source of these ideas is provided by results of full numerical integrations. The asymptotics can provide interpretations of the results that would not be evident from the numerics themselves. Thus, there are symbiotic interactions between numerics and asymptotics.

With large numbers of reactions, systematic procedures are needed for achieving simplifications. An important principle that can be established for excluding reaction steps is a comparison principle[10]. An accuracy measure is selected, and a reactive whose omission produces a change less than the accuracy measure is deemed excludable. The comparison principle can be defined relatively straightforwardly once a short asymptotic description is available, and it can be used to test whether additional, new, reaction steps should be added. There is a hazard here in that, in principle, one step may not be very important, but a large number of relatively unimportant steps may have an appreciable cumulative influence. Nevertheless, comparison principles offer useful working hypothesis of achieving simplifications.

Aside from omitting reaction steps outright, there are two types of systematic chemical-kinetic approximations for achieving simplifications – steady-state and partial-equilibrium approximations – the first for chemical species and the second for reaction steps. Although they are both aspects of the same, more general, type of reduction of order of differential systems, they have been treated separately to facilitate interpretation. Their most important attribute is that well-defined methods exist to test their applicability, again subject to a specified accuracy measure. Introduction of steady-state and partial-equilibrium approximations reduces the number of independent reaction steps. The reduced system can then be employed as a comprehensible basis for introduction of asymptotic approximations, or, alternatively, full numerical integrations can be employed with the reduced mechanism. Both of these procedures have been used for investigating hydrocarbon flames. To be honest, it must be admitted that, at the present, rapidly developing, stage of analysis, accuracy checks of the approximations are not always made.

Reductions to one-step mechanisms have not proven successful for hydrocarbon flames. At least two steps are needed to retain correct qualitative features, and three steps or four steps must be employed for reasonable accuracy. It can be considered surprising that, when starting with more than 200 steps, only three or four suitably defined steps can provide an acceptable description. Yet, this has been demonstrated[10] for stoichiometric methane-air flames with 300 K $\leq$ It $\leq$ 500 K and 1 atm $\leq$ P $\leq$ 20 atm, and it seems likely to apply for lean methane-air flames as well. Evaluations are more difficult for rich flames because of uncertainties in elementary steps and in their rates for fuel-containing species; even full numerical integrations exhibit discrepancies with experiment for rich flames.

For the methane flame (overall $CH_4 + 2O_2 \rightarrow CO_2 + 2H_2O$) the two-step mechanism is $CH_4 + O_2 \rightarrow aCO + bH_2 + cH_2O + dCO_2$, with a, b, c and d determined by steady states for radicals and by partial equilibrium for the water-gas shift, $CO + H_2O \rightleftarrows CO_2 + H_2$, followed by oxidation of CO and H_2, $aCO + bH_2 + cH_2O + dCO_2 + O_2 \rightarrow CO_2 + 2H_2O$. The first step proceeds at the rate of $CH_4 + H \rightarrow CH_3 + H_2$, and the second at the rate of $H + O_2 + M \rightarrow HO_2 + M$. The first step occurs in a thin fuel-consumption layer, located at a temperature where the geometric mean of the rates of $CH_4 + H \rightarrow CH_3 + H_2$ and $H + O_2 + M \rightarrow HO_2 + M$ (both of which lead to radical removal) balances the rate of $H + O_2 \rightarrow OH + O$, the main branching step. The second step occurs immediately downstream from this in a somewhat thicker layer. The small parameter of expansion of the asymptotic analysis is a ratio of the rates of the branching and fuel-consumption steps, and therefore the analysis involves RRA; AEA plays no role whatever[10]. Analyses of stability and response for the two-step mechanism are needed because the results may differ from those obtained previously with AEA.

In the three-step mechanism for the premixed methane flame, the reversible water-gas step departs from equilibrium. This gives rise to a water-gas nonequilibrium layer between the two just identified. Still only RRA is involved[10], and reasonably good agreement for burning velocities is obtained if account is taken of elementary steps in the fuel-consumption zone that are found to be important by the comparison principle. The stability and responses probably would be much like that for the two-step mechanism; it might be different for the four-step mechanism, which relaxes the steady state for H and thereby brings the rate of $H + O_2 \rightarrow OH + O$ into the fuel-zone analysis.

With all of the fuel chemistry confined to the narrow fuel-consumption zone, large numbers of fuel-chemistry steps are readily included in the asymptotic ananlysis[10]. This enables the same RRA approach to be employed for other hydrocarbon fuels besides methane. The same kinds of ideas have even been applied to methanol flames, for example; here it seems that a five-step mechanism may be needed[11]. Recent studies[12] of flames employing mixtures of H_2 and CO as fuels suggest that related RRA methods may work, although the structures are not just those of the methane flame with the fuel-consumption zone removed (and replaced by a cut-off at the cross-over temperature of $H + O_2 \rightarrow OH + O$ and $H + O_2 + M \rightarrow HO_2 + M$). Thus, there is great activity in asymptotic analyses of premixed flames with detailed chemistry; much remains to be done, and many things are likely to be learned in the near future.

5. HYDROGEN-HALOGEN DIFFUSION FLAMES

A general question of interest in diffusion flames concerns whether the kinetic mechanisms are the same as those for premixed flames[1]. Of course, complete reaction schemes for diffusion flames involve the same reaction steps as the premixed flame of the same reactants. however, the same step may have a different relative importance in the premixed and diffusion flames (it may be negligible in one but not in the other, for example), or the step may occur at different places in the two flames, or the best reduced mechanisms may differ. In these cases, the simplified description of the mechanisms would be different for the two flames. Questions of whether this happens are only beginning to be explored.

Hydrogen-halogen flames provide an example of the possibility. In the premixed flame, the chain-carrying steps $Ha + H_2 \rightarrow H + HHa$ and $H + Ha_2 \rightarrow Ha + HHa$ occur at the same position. In the different flame it is conceivable[13] that the first of these may occur at a point near the H_2 side, and the H atoms thereby released may diffuse toward the Ha_2 side, to a different position where the second step occurs, releasing Ha that diffuses toward the H_2 side to support the first step. Thus, the two steps may be separated spatially. Although it is easy to invent reaction-sheet models with this two-sheet structure, it is not known whether such models can ever be realistic for any hydrogen-halogen flames. In addition, general criteria for this kind of splitting to occur have not been defined. Hydrogen-halogen diffusion flames are in need of further investigation.

6. METHANE-AIR DIFFUSION FLAMES

Recent study has been given to the structure of methane-air diffusion flames by asymptotic methods[14,15] but unresolved conflicts remain. In one view[14], the structure entirely parallels that of the premixed flame, with the fuel-consumption zone occurring on the fuel side of the stoichiometric mixture fraction and the zone of H_2 and CO combustion (and oxygen consumption) extending from there to the lean side of stoichiometry. Water-gas nonequilibrium exists on the rich side of the latter zone in this view. Thus, moving from upstream to downstream in the premixed flame corresponds to moving from rich to lean in the diffusion flame. This model successfully predicts oxygen leakage to the fuel side and strain rates for extinction (perhaps with some inaccuracy). To the extent that the model is correct, the premixed-flame and diffusion-flame structures and kinetic mechanisms are essentially the same.

As an alternative view[15] of the diffusion-flame structure, it has been proposed that the fuel may be consumed by radicals not in partial equilibrium. On the rich side of stoichiometry there may be a diffusion flame within the diffusion flame, where radicals diffusion from the lean side consume the fuel diffusing from the rich side. On the extreme lean side, at a temperature where the rates of $H + O_2 \rightarrow OH + O$ and $H + O_2 + M \rightarrow HO_2 + M$ are equal, the radicals are removed by three-body processes; between this position and the fuel-consumption zone, oxygen is consumed and radicals produced through finite rates of $H + O_2 \rightleftharpoons OH + O$, with the other hydrogen-oxygen steps impartial equilibrium. This last step also maintains partial equilibrium in a region within the lean part of this radical-production region, separated at a "sudden-freezing" point from a rich-part region in which the reverse of this last stop can be neglected. This model may predict extinction through finite-rate chemistry in the radical-production or fuel-consumption zone, the latter being described by AEA (while the former involves RRA).

If this second view is correct, then there are a number of ways in which the diffusion-flame kinetics differ from the premixed-flame kinetics; e.g., AEA plays a role. More study is needed to identify which one of these two alternative models is the better, and whether a different description may be even better. Further investigations may help to clarify relationships between premixed-flame and diffusion-flame kinetics.

7. CONCLUDING REMARKS

From these discussions it is seen that a variety of new asymptotic methods have become available for describing structures and behaviors of premixed and diffusion flames, with detailed chemistry taken into account. With systematic methods, complex chemistry having more than 200 steps can readily be addressed by asymptotics. Intensive study along these lines is in process, and rapid discoveries of new ideas about flame structures are anticipated in the near future. Many different problems remain to be addressed, and there is need for even greater activity in the area.

REFERENCES

1. Williams, F.A.: Influences of Detailed Chemistry on Asymptotic Approximations for Flame Structure, Proceedings of the Conference on Mathematical Modeling in Combustion, Lyon, France, April 1987, NATO ASI Series, Martinus Nijhoff Publ., to appear.

2. Williams, F.A.: Combustion Theory. 2nd ed., Benjamin/Cummings, Menlo Park, CA, 1985.

3. Gardner, W.C., Jr.: Combustion Chemistry. Springer Verlag, New York, 1984.

4. Rogg, B. and Wichman, I.S.: Combust. Flame 62, 271 (1985).

5. Linan, A. and Rodgriguez, M.: Combustion and Non Linear Phenomena. Clavin, P., Larrouturou B. and Pelce, P., eds., Les Editions de Physique, le Ulis, 1986, p. 51.

6. Rogg, B., Linan, A. and Williams, F.A.: Combust. Flame 65, 79 (1986).

7. Rogg, B.: Combust. Flame 65, 113 (1986).

8. Mikolaitis, D.W.: unpublished, 1987.

9. Rogg, B.: Comb. Sci, and Tech. 45, 317 (1985).

10. Peters, N. and Williams, F.A.: Combust. Flame, 68, 1985 (1987).

11. Seshadri, K.: unpublished, 1987.

12. Rogg, B. and Williams, F.A.: unpublished, 1987.

13. Zebib A., Williams, F.A. and Kassoy, D.R.,: Combust. Sci. and Tech. <u>10</u>, 37 (1975).

14. Seshadri, K. and Peters, N.: Asymptotic Structure and Extinction of Methane-Air Diffusion Flames. Submitted to Combst. and Flame, 1987.

15. Trevino, C. and Williams, F.A.: An Asymptotic Analysis of the Structure and Extinction of Methane-Air Diffusion Flames. In preparation, 1987.

UNSTEADY TRANSITION FROM SUB- TO SUPERCRITICAL EVAPORATION REGIME

Akira Umemura
Department of Mechanical Engineering, Yamagata University
Jonan 4-3-16, Yonezawa, 992 JAPAN

Introduction

Knowledge of droplet combustion at elevated pressures and temperatures
is important for design of high-output combustors for Diesel engines
and liquid propellant rocket motors. This is why extensive experimental
and theoretical investigations have been performed for the elementary
problem of single droplet combustion in stagnant environments [1-10].
There, however, remains a fundamental question about the transition
from sub- to supercritical evaporation regime, which may be encoutered
when the ambient pressure and temperature are greater than the thermodynamic critical pressure and temperature of the fuel.

If a droplet heats through its critical temperature, it will no
longer have a discontinuous liquid-gas interface. Instead, phase change
takes place in a spatially continuous way and the subsequent combustion
rate is controlled not by evaporation rate but the diffusion of fuel
vapor. This is a situation which Spalding [2] analized assuming that
the droplet burns like an initially well defined puff of fuel vapor.
The analysis is valid for such an extreme case that the lapse of time
before the droplet surface reaches the critical state is much smaller
than the entire combustion time. There is another approach to apply
a low-pressure quasi-steady droplet combustion model [11]. Since, in
some condition, the droplet may keep the surface throughout its lifetime, this approach also has a firm physical basis. But it deals with
the other extreme case.

So far there is no established theory which enables us to treat
a more realistic situation in which the droplet surface reaches the
critical state at a finite radius. In addition to its unsteady nature
the phenomenon at the transition stage is complicated by uncertainties
in the determination of thermodynamic and transport properties which
may exhibit anomalous behavior near the critical point. Recently a
new mathematical model was developed by the author [12], which emphasizes
the importance of vanishing diffusion coeffiscient at the critical
liquid-gas interface for the successful analysis of combusion
characteritics at high pressures. Since it was based on self-similar

solutions, the analysis however could not describe the transition itself.

This investgation is on the course of extension from the previous one and aims at obtaining a basic understanding of the physical mechanism of unsteady transition from sub- to supercritical evaporation regime. Although the effect of thermodynamic fluctuations might become significant near the critical point, we confine ourselves in the realm of the classical continuum theory which assumes the validity of local-equilibrium hypothsis and of linear phenomenological theory [13]. Further, when there is a liquid-gas interface, we assume the hold of phase equilibrium which is consistent with the local-equilibrium hypothesis.

State-Space Description

The characteristic evaporation feature at supercritical pressures will be well described in the state space introduced below.

We consider the spherically symmetric, isobaric evaporation process of a single fuel droplet which is suddenly immersed in an otherwise quiescent inert gas with uniform temperature T_g. The droplet has a surface of radius X_1 at an early time t_1. The curves $a_1's_1'b'$ and $a_1''s_1''b''$ in Fig.1(a) show the temperature distribution $T(t,x)$ and the fuel concentration distribution $Y(t,x)$ at time $t = t_1$. The temperature is continuous across the droplet surface while the concentration becomes discontinuous and has the two interfacial values Y_{s1} and Y_{*s1} which are determined from the surface temperature T_{s1} through the phase equilibrium relations relevant at the prescribed pressure P. Note that both $T(t_1,x)$ and $Y(t_1,x)$ are functions of x in either side of the liquid and gas phases. Eliminating x from $T(t_1,x)$ and $Y(t_1,x)$, we obtain the curve $A_1S_{L1}S_{G1}B$ in Fig.1(b). This expresses the instantaneous fuel concentration Y as a function of temperature T. A similar graph of $Y(t,T)$ is obtained at another time $t = t_2 > t_1$.

With the lapse of time the droplet heats up and dissolves the ambient gas inside by diffusion, so that the point expressing the thermodynamic state at the droplet center $x = 0$ moves from A_1 to A_2 in Fig.1(b). At the same time the two points which correspond to the state at the liquid-gas interface move along the isobaric saturation curve $S_{L1}S_{L2}CS_{G2}S_{G1}$ toward the point C which expresses the thermodynamic critical state of the fuel-inert gas binary mixture at the given pressure. For the case when the droplet has the surface throughout its lifetime, the locus of the central state point A terminates at a point on the saturated liquid curve $S_{L1}S_{L2}C$. Otherwise, the transition from sub- to supercritical evaporation regime takes place during the droplet lifetime. The critical case is when the fuel concentration profile Y(t,T) passes through the critical point C in Fig.1(b), i.e., when the state at the droplet surface just coincides with the mixture critical state. After the transition, there is no liquid-gas interface and the temperature and fuel concentration distributions become smoothly continuous everwhere as shown by the curves $a_3'b'$, $a_3''b''$ and A_3B.

Whether the droplet experiences the transition during its lifetime depends not only on the pressure P but also on other system parameters such as the ambient gas temperature T_g and the initial liquid fuel temperature T_1. Thus, we can distingush three stages for a series of realizations when one of the system parameters P, T_g and T_1 is increased with the others fixed. At small values of the parameter the droplet has the liquid-gas interface throughout its whole lifetime (Stage I). The first transition from sub- to supercritical evaporation regime is observed in such a situation that the droplet surface becomes the critical state just when the droplet is about to evaporate completely (Stage II). For greater values of the parameter the transition occurs at a finite droplet radius (Stage III). In this paper we are interested in the last case.

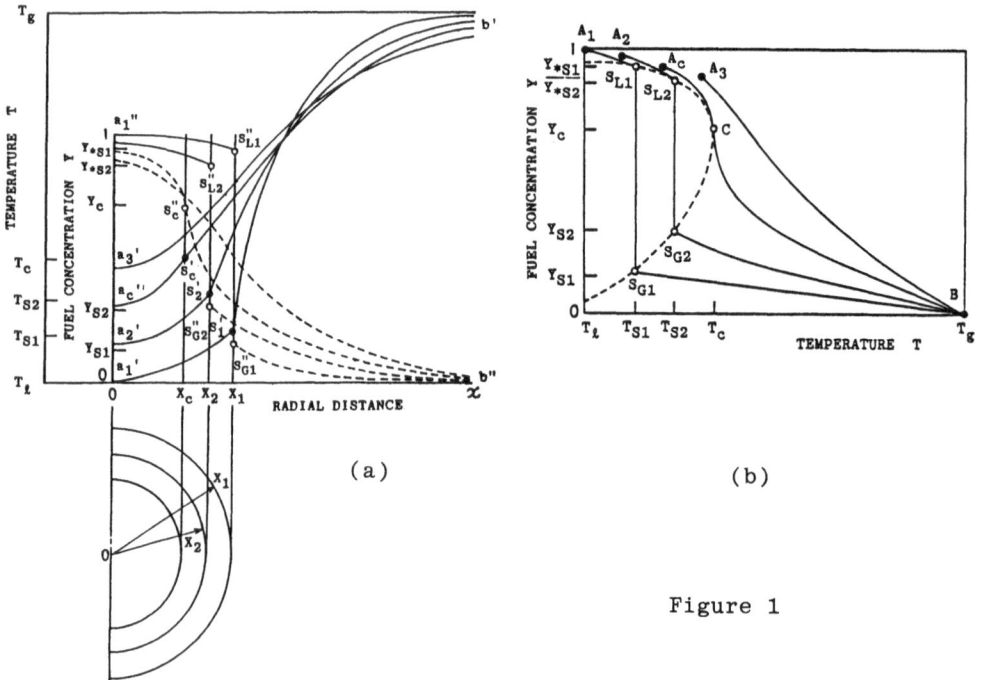

(a)

(b)

Figure 1

Plane One-Dimensional Evaporation Problem

As a plane one-dimensional version of the spherically symmetric droplet evaporation problem, we consider the plane-symmetric evaporation of a pure liquid fuel slab with a finite initial thickness $2X_0$ and temperature T_1. The undisturbed ambient inert gas has a uniform temperature $T_g > T_c$. The evaporation is assumed to proceed at a constant pressure P greater than the critical pressure of the fuel, P_c, so that the phenomenon is described by the temperature field $T(t,x)$ and the fuel mass fraction field $Y(t,T)$ at each time t, where x is measured from the plane of symmetry of the slab. Although this problem lacks the effect of curvature, the other essential properties are all involved.

The phenomenon is governed by the following transport equations.

$$\frac{\partial \rho}{\partial t} + \frac{\partial \rho u}{\partial x} = 0 \tag{1}$$

$$\rho [\frac{\partial h}{\partial t} + u \frac{\partial h}{\partial x}] = \frac{\partial}{\partial x} \lambda \frac{\partial T}{\partial x} \tag{2}$$

$$\rho [\frac{\partial Y}{\partial t} + u \frac{\partial Y}{\partial x}] = \frac{\partial}{\partial x} \rho D \frac{\partial Y}{\partial x} \tag{3}$$

where ρ, u and h denote the mixture mass density, velocity and enthalpy, respectively. The equations must be suplemented with a state equation $\rho = \rho(T,Y;P)$, a caloric equation $h = h(T,Y;P)$ and appropriate expressions for the mixture thermal conductivity λ and binary diffusion coefficient D as functions of T, Y and P. The equations are applicable to both phases. We shall use subscript (*) to distinguish liquid-phase properties.

Basic Equations

The relevant governing equation system in the state space can be derived from the above field equations by introducing a new variable

$$\Psi(t,T) = \lambda \frac{\partial T}{\partial x}$$

corresponding to the negative conductive heat flux. Regarding Y as a function of t and T, Eq.(3) is combined with Eq.(2) to yield

$$\frac{h_T \rho \lambda}{c_p \Psi} (\frac{\partial Y}{\partial t})_T = \frac{\partial}{\partial T} \frac{\rho D}{\lambda} \Psi \frac{\partial Y}{\partial T} - \frac{1}{c_p} \frac{\partial Y}{\partial T} \frac{\partial \Psi}{\partial T} \tag{4}$$

where we have defined the modified specific heat at constant pressure by

$$c_p \triangleq h_T + h_Y \frac{\partial Y}{\partial T} \tag{5}$$

Equation (4) governs the time evolution of the concentration profile Y(t,T) in the state space of Fig.1(b). The governing equation for Ψ is obtained by differentiating Eq.(2) with respect to x and using Eqs.(1) and (3) to eliminate the velocity u.

$$\frac{\rho\lambda}{\Psi^2} \frac{\partial\Psi}{\partial t} - (\frac{\rho_Y}{\rho} + \frac{\lambda_Y}{\lambda}) \frac{\rho\lambda}{\Psi} \frac{\partial Y}{\partial t} = \frac{\partial}{\partial T} [\frac{1}{c_p} \frac{\partial\Psi}{\partial T} - \frac{h_Y\rho\lambda}{c_p\Psi} \frac{\partial Y}{\partial t}] \tag{6}$$

Equations (4) and (6) present the basic equations which govern heat and mass transfers in the state space. We note the following conservative property. Equation (4) is transformed into

$$\frac{\rho\lambda}{\Psi} \frac{\partial Y}{\partial t} + [\frac{1}{c_p} \frac{\partial\Psi}{\partial T} - \frac{h_Y}{c_p} \frac{\rho\lambda}{\Psi} \frac{\partial Y}{\partial t}] \frac{\partial Y}{\partial T} = \frac{\partial}{\partial T} \frac{\rho D}{\lambda} \Psi \frac{\partial Y}{\partial T}$$

If we regard the coefficients of $\partial Y/\partial t$ and $\partial Y/\partial T$ as the "mass density" and "mass flux" in the T-space, then the continuity equation becomes

$$\frac{\partial}{\partial t} \frac{\rho\lambda}{\Psi} + \frac{\partial}{\partial T} [\frac{1}{c_p} \frac{\partial\Psi}{\partial T} - \frac{h_Y}{c_p} \frac{\rho\lambda}{\Psi} \frac{\partial Y}{\partial t}] = 0$$

This is just the heat transport equation (4), so that the quantities $\rho\lambda/\Psi$ and Y are conservative in the T-space. Correspondingly, the diffusion coefficient in the T-space is given by $(\Psi^2/\lambda^2)D$.

Equations (4) and (6) are subject to the following initial and boundary conditions.

At initial time t = 0: $\Psi_* = \Psi = 0$, $Y_* = 1$, $Y = 0$

At the plane of symmetry $T = T_0(t)$:

$$\Psi_* = 0, \quad Y_* = Y_{*0}(t)$$

$$\frac{dT_0}{dt} = \frac{1}{2\rho_*\lambda_*h_{T*}} [1 - h_{Y*} \frac{\rho_*D_*}{\lambda} \frac{\partial Y_*}{\partial T}] \frac{\partial\Psi_*}{\partial T}^2 \tag{7}$$

$$\frac{dY_{*0}}{dt} = \frac{D_*}{2\lambda_*^2} \frac{\partial Y_*}{\partial T} \frac{\partial\Psi_*}{\partial T}^2 \tag{8}$$

At the liquid-gas interface $T = T_S(t)$:

$$\Psi_* = \Psi_{*S}(t), \quad \Psi = \Psi_S(t), \quad Y_* = Y_{*S}(t), \quad Y = Y_S(t)$$

$$\mu_*^{(i)}(T_S,Y_{*S},Y_S;P) = \mu^{(i)}(T_S,Y_{*S},Y_S;P) \qquad (i = F, I) \tag{9}$$

$$M_S = \frac{1}{c_{p*}} \frac{\partial \Psi_*}{\partial T} - \frac{\rho_* \lambda_*}{c_{p*} \Psi_*} \left(h_{T*} \frac{dT_S}{dt} + h_{Y*} \frac{dY_{*S}}{dt} \right)$$

$$= \frac{1}{c_p} \frac{\partial \Psi}{\partial T} - \frac{\rho \lambda}{c_p \Psi} \left(h_T \frac{dT_S}{dt} + h_Y \frac{dY_S}{dt} \right) \tag{10}$$

$$\Psi - \Psi_* = M_S (h_S - h_{*S}) \tag{11}$$

$$M_e = M_S Y_* - \frac{\rho_* D_*}{\lambda_*} \Psi_* \frac{\partial Y_*}{\partial T} = M_S Y - \frac{\rho D}{\lambda} \Psi \frac{\partial Y}{\partial T} \tag{12}$$

At infinity $T = T_g$: $Y = 0$

   The equation system above constitutes the Stefan problem in the
state space, where each boundary except infinity moves with time
according to Eqs.(7) and (10) which are derived from conservation
of the overall mass flux across the corresponding physical boundaries.
The mass fraction of fuel at the plane of symmetry, Y_{*0}, is determined
from Eq.(8). Equation (9) provides the determining equations for Y_{*S}
and Y_S, and Eqs.(11) and (12), for Ψ_{*S} and Ψ_S. The quantity M_e
expresses the fuel mass evaporation rate.
   Important information about the transition is immediately drawn
from Fig.1(b). Since the tangent to the saturation curve at the critical
point becomes vertical, the derivative of the concentration profile,
$\partial Y / \partial T$, at the transition time must diverge at the critical point.
This divergence requires the vanishing binary diffusion coefficient
at the critical interface, so that the transition might take place
[12]. As an expression for the diffusion coefficient which satisfies
this condition, we may use

$$D = \alpha D_0 , \quad \alpha = \frac{Y}{RT} \frac{\partial \mu}{\partial Y} \cong \frac{Y_c}{2RT_c} \frac{\partial^3 \mu}{\partial Y^3}\bigg|_c (Y - Y_c)^2 \tag{13}$$

which can be derived invoking the Brownian motion theory.
   Along with the above formulus, in the later analysis we need formulae
which explicity express the interfacial states about the critical point.
The satuation curve near the critical point can be approximated by
a parabolic curve and we obtain the following expression from the
consideration of the phase equilibrium relations.

$$Y_{*S} - Y_c = Y_c - Y_S = \sqrt{\frac{2}{K}} \sqrt{T_c - T_S} \tag{14}$$

with

$$\frac{2}{K} = \frac{6 \partial^2 \mu / \partial T \partial Y}{\partial^3 \mu / \partial Y^3}\bigg|_c$$

The latent heat of vaporization is expressible in the form

$$L = h(T_S, Y_S) - h_*(T_S, Y_{*S}) = 2(-h_{Y_C})|Y_S - Y_C| \tag{15}$$

The behavior of the surface temperature changes depending on the system pressure value. This is easy to see in the state space. For the subcritical pressure case $P > P_C$, the isobaric saturation curve has a boiling point of the fuel ($Y_S = Y_{*S} = 1$ at $T_S = T_b$). This gives an upper limit for the attainable surface temperature. While both the boiling point and the critical point divide the saturation curve into the saturated liquid and vapor branches, the former differs from the latter in that the latent heat of vaporization, L, has a non-zero value at the boiling value. Since $T_g > T_C > T_b$, the slab with finite thickness always receives a finite heat flux Ψ_S from the gas phase at the surface. If $\Psi_S > M_S L$, the liquid is heated up to accommodate the surface temperature so that the condition $\Psi_S = M_S L$ can attain finally. It is apparent that a non-zero L is necessary for the wet-bulb condition to realize. For the supercritical pressure case, on the other hand, there is no upper limit for the surface temperature in such a sence as mentioned above. Even in this case the slab may attain the wet-bulb condition when it evaporates completely before the surface reaches the critical state. As the surface temperature approaches the critical temperature, the latent heat L reduces to zero. Therefore, the heat transferred from the gas phase, which does not vary greatly with time, is used to heat up the liquid phase and the surface can pass through the critical condition. This situation is nearly the same as the pure heat transport problem with spatially varying properties, suggesting a finite rate of change in the surface temperature at the time of transition from sub- to supercritical evaporation regime.

Unsteady Transition

Suppose that, in a realization at prescribed P, T_g and T_1, the surface of the liquid slab reached the critical condition ($T_S = T_C$, $Y_S = Y_{*S} = Y_C$) at time $t = t_C$. To analyze this transient process from sub- to supercritical evaporation regime, we focus attention on the behaviors of $Y(t, T)$ in the vicinity of the moving liquid-gas interface during a short time period, $t_C - t_r \Delta t \le t \le t_C + t_r \Delta t$. In the state space of Fig.1 we consider a narrow vertical zone, $T_C - T_C \Delta T \le T \le T_C + T_C \Delta T$, where the instantaneous concentration profile $Y(t, T)$ changes abruptly with T. Denoting by ΔY the order of magnitude of this change, the gradient $\partial Y/\partial T$ is estimated to be of the order $\Delta Y/\Delta T$, which must

be great because of the divergence nature of the critical concentration
profile. The asymptotic analysis developed in what follows is based
on this secular behavior of the concentration profile around the
critical point.

We stretch the variables T, t and Y by

$$\theta = \frac{T - T_c}{T_c \Delta T} \quad , \quad \tau = \frac{t - t_c}{\frac{\rho g h T g X_0}{\lambda_g} \Delta t} \quad , \quad \eta = \frac{Y - Y_c}{\Delta Y}$$

while, for the conductive heat flux, we put

$$\psi = \frac{\lambda_g (T_g - T_c)}{X_0} [\Psi_c + \Delta \Psi \phi]$$

For the transition at Stage III the normalized conductive heat flux
at the critical interface, Ψ_c, is of $O(1)$. That is, Ψ_c takes a value of
the same order as Ψ_s and Ψ_{*s} before the transition period, so that the
temperature within the liquid slab increases with time in the same
way as before and after the transition period.

The scales ΔT and ΔY are chosen as follows. Using the stretched
variables we first express the saturation curve in the form

$$-\eta_s = \eta_{*s} = \epsilon \sqrt{-\theta_s}$$

where ϵ is a small parameter. This is derived by putting

$$\epsilon^2 \frac{\Delta Y^2}{\Delta T} = \frac{6T_c}{\partial^3 \mu / \partial Y^3} \left. \frac{\partial^2 \mu / \partial T \partial Y}{} \right|_c \tag{16}$$

and indicates that the values of interfacial concentrations become $O(\epsilon)$
when measured on the scale ΔY. The meaning of the parameter ϵ is
understandable when we consider the difference in form between the
concentration profile and the saturation curve. According to the
previous self-similar analysis [12], the critical concentration profile
η changes like $(\theta)^{1/3}$. This would be the case in the present problem,
too. Since the saturation curve, on the other hand, changes like
$(-\theta)^{1/2}$, it follows that, when ΔY is chosen so as to normalize the
concentration profile, the stretched variable η takes small values
on the staturation curve. The smaller ΔT is taken, the smaller the
value of η_s and η_{*s} become. Thus, the parameter ϵ can be regarded as a
measure which expresses a "distance" from the critical point. As an
example Fig.2 shows the pressure dependences of $T_c \partial^2 \mu / \partial T \partial Y |_c$ and
$\partial^3 \mu / \partial Y^3 |_c$ for the normal butane-nitrogen binary system. They were
calculated on the basis of the Redlich-Kwong equation of state with
appropriate mixing rules. From the figure the quantity on the right-
hand side of Eq.(16) is found to take a value of $O(1)$ except near

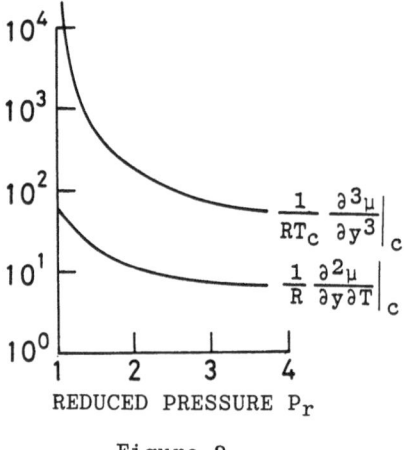

Figure 2

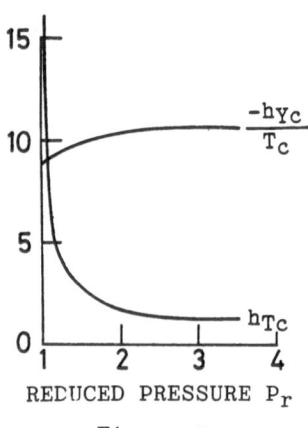

Figure 3

$P_r = 1$. We may except similar property for other binary component systems.

The thermodynamic corrector α assumes

$$\alpha = \frac{Y_c}{2RT_c} \frac{\partial^3 \mu}{\partial Y^3}\bigg|_c \Delta Y^2 [\eta^2 + \frac{1}{3} \epsilon^2 \theta] \tag{17}$$

Taking into account Eqs.(13), (15) and (17) the boundary conditions (11) and (12) are combined to yield

$$\frac{(-h_{Yc})\rho_c D_{0c}}{T_c \lambda_c} \frac{Y_c}{2RT_c} \frac{\partial^3 \mu}{\partial Y^3}\bigg|_c \frac{\Delta Y^3}{\Delta T} [-(\eta^2 + \frac{1}{3} \epsilon^2 \theta) \frac{\partial \eta}{\partial \theta}]_{\theta_s \to 0} = 1$$

for the critical interface, so that it is natural to choose ΔT and ΔY to satisfy the condition

$$\frac{\Delta T}{\Delta Y^3} = \frac{(-h_{Yc})\rho_c D_{0c}}{T_c \lambda_c} \frac{Y_c}{2RT_c} \frac{\partial^3 \mu}{\partial Y^3}\bigg|_c \tag{18}$$

From Eqs.(19) and (22) we obtain

$$\Delta Y = \epsilon^2 / \frac{3(-h_{Yc})\rho_c D_{0c} Y_c}{RT_c \lambda_c} \frac{\partial^2 \mu}{\partial T \partial Y}\bigg|_c \tag{19}$$

$$\Delta T = \epsilon^6 \frac{\partial^3 \mu}{\partial Y^3}\bigg|_c / \frac{54(-h_{Yc})^2 \rho_c^2 D_{0c}^2 Y_c^2}{R^2 T_c \lambda_c^2} (\frac{\partial^2 \mu}{\partial T \partial Y}\bigg|_c)^3 \tag{20}$$

Figure 3 shows the pressure dependences of h_{Tc} and $(-h_{Yc})/T_c$ for the same binary component system as before. Both quantities take finite values unless $Y = 1$ or 0. Therefore, the modified specific heat c_p can be represented with its diverging term $h_Y \partial Y/\partial T$. This approximation is justified if the condition

$$\varepsilon^4 \; << \; \frac{18(-h_{Yc})^2 \rho_c D_{0c} Y_c}{RT_c h_{Tc} \lambda_c} \; \frac{(\partial^2 \mu / \partial T \partial Y|_c)^2}{\partial^3 \mu / \partial Y^3|_c} \tag{21}$$

is satisfied. In view of Fig.2 and 3, the quantity on the right-hand side takes a value of $O(1)$ or larger except near $P_r = 1$. Hence, the parameter ε should be chosen so small that the ΔY calculated from Eq.(19) becomes much smaller than unity. This choice satisfies the requirement $\Delta T << 1$ automatically. For the case of $P_r \simeq 1$, Eq.(21) controls the choice of ε. In this case the transition from sub- to supercritical evaporation regime, however, hardly takes place during the lifetime of the liquid slab; The slab will evaporate completely before the surface reaches the critical condition.

The other material properties which have a non-zero critical value are also expanded into the Taylor series in T and Y about the critical point. Substituting the expansions into Eqs.(4)-(6) and (10)-(12) we obtain their perturbation equations with respect to the small parameters ΔT, ΔY, $\Delta \Psi$ and Δt. It is found from the interfacial conditions (11) and (12) that the overall mass flux across the liquid-gas interface should be normalized according to

$$\hat{M}_s \; = \; \frac{M_s X_0 (-h_{Yc})}{\lambda_g (T_g - T_c)} \; \frac{\Delta Y}{\Delta \Psi} \tag{22}$$

Hence, we have, in place of Eq.(11),

$$\phi_s - \phi_{*s} = \hat{M}_s (\eta_{*s} - \eta_s) = \hat{M}_s \cdot 2\varepsilon \sqrt{-\theta_s} \tag{23}$$

This equation does not necessarily mean that ϕ_s and ϕ_{*s} are quantities of $O(\varepsilon)$. Instead they may share a value of $O(1)$, ϕ_c, say, but their difference should be of $O(\varepsilon)$.

The other interfacial conditions and the governing differential equations assume the following form in the lowest order of approximation. From Eq.(6) we have the relation

$$\Psi_c \; \frac{\Delta \Psi \Delta t}{\Delta Y \Delta T} = \frac{\rho_c \lambda_c (-h_{Yc})}{\rho_g \lambda_g h_{Tc} T_c} \; (\frac{T_c}{T_g - T_c})^2 \tag{24}$$

which casts Eqs.(11) and (13) into the form

$$\frac{\partial}{\partial \theta} \; (\eta^2 + \frac{1}{3} \varepsilon^2 \theta) \frac{\partial \eta}{\partial \theta} = 0 \tag{25}$$

$$\frac{\partial}{\partial \theta} [\frac{\partial \phi / \partial \theta + \partial \eta / \partial \tau}{\partial \eta / \partial \theta}] = 0 \tag{26}$$

The interfacial conditions (11) and (12) become

$$\hat{M}_s = -\frac{\partial\phi/\partial\theta\,|_s}{\partial\eta/\partial\theta\,|_s} - \frac{\partial\eta_s/\partial\tau}{\partial\eta/\partial\theta\,|_s} = \frac{\partial\phi_*/\partial\theta\,|_s}{\partial\eta_*/\partial\theta\,|_s} - \frac{\partial\eta_{*s}/\partial\tau}{\partial\eta/\partial\theta\,|_s} \qquad (27)$$

$$M_e = \frac{M_e}{\psi_c \dfrac{\lambda g(T_g - T_c)}{(-h\gamma_c)X_0}} - \frac{\gamma_c\Delta\Psi}{\Psi_c\Delta Y}\,\hat{M}_s = -\eta_s{}^2\,\frac{\partial\eta}{\partial\theta}\bigg|_s = -\eta_{*s}{}^2\,\frac{\partial\eta_*}{\partial\theta}\bigg|_s \qquad (28)$$

In Eq.(25) we have included a higher term for the following reason. Although negiligible almost in the whole domain relevant, the contribution of the term $\varepsilon^2\theta/3$ becomes important in such a region as $\theta \geq 3\eta^2/\varepsilon^2$. For $\tau \leq 0$ when possible concentration profiles do not cross this region, we can neglect it. However, for $\tau > o$ its effect becomes significant around the intersecting point of the concentration profile with the θ-axis. In fact, we shall see that this term plays an essential role to produce a gradual change in the concentration profile which corresponds to the spatially continuous phase change. Since the material properties change with concentration much greater than with temperature, some careful treatment is needed to derive Eq.(26) from Eq.(6). The second term on the left-hand side of Eq.(6) is canceled by a part of the second term on the right-hand side due to the concentration dependences of ρ and λ. Equations (25) and (26), subject to the boundary conditions (23), (27) and (28), are easily solved to yield

for $\tau \leq 0$

$$\theta - \theta_s = -\frac{1}{3\hat{M}_e}\,[\eta^3 + \varepsilon^3(-\theta_s)^{\frac{3}{2}}] = -\frac{1}{3\hat{M}_e}\,[\eta_*{}^3 - \varepsilon^3(-\theta_s)^{\frac{3}{2}}] \qquad (29)$$

$$\begin{aligned}
\phi - \phi_0 - \hat{M}_s\varepsilon\sqrt{-\theta_s} &= -\frac{1}{12}\left(\frac{d}{d\tau}\,\frac{1}{\hat{M}_e}\right)(\eta^4 - \varepsilon^4\theta_s{}^2) \\
&\quad - [\hat{M}_s + \frac{1}{3}\,\varepsilon^3\,\frac{d}{d\tau}\,\frac{(-\theta_s)^{\frac{3}{2}}}{\hat{M}_e}](\eta + \varepsilon\sqrt{-\theta_s}) \\[2mm]
\phi - \phi_0 + \hat{M}_s\varepsilon\sqrt{-\theta_s} &= -\frac{1}{12}\left(\frac{d}{d\tau}\,\frac{1}{\hat{M}_e}\right)(\eta^4 - \varepsilon^4\theta_s{}^2) \\
&\quad - [\hat{M}_s - \frac{1}{3}\,\varepsilon^3\,\frac{d}{d\tau}\,\frac{(-\theta_s)^{\frac{3}{2}}}{\hat{M}_e}](\eta_* - \varepsilon\sqrt{-\theta_s})
\end{aligned}\right\} \qquad (30)$$

and for $\tau \geq 0$

$$\theta = (\theta_0 + \frac{54m_e{}^2}{\varepsilon^6})\,\exp(-\frac{\varepsilon^2}{3m_e}\eta) - \frac{54m_e{}^2}{\varepsilon^6} + \frac{18m_e}{\varepsilon^4}\,\eta - \frac{3}{\varepsilon^3}\,\eta^2 \qquad (31)$$

$$\begin{aligned}
\phi - \phi_0 &= \frac{9}{\varepsilon^4}\,\frac{dm_e}{d\tau}\,\eta^2 - [(\frac{\theta_0}{m_e} + \frac{54m_e}{\varepsilon^6})\,\frac{dm_e}{d\tau}\,\exp(-\frac{\varepsilon^2}{3m_e}\eta) \\
&\quad + (m_s + \frac{54}{\varepsilon^6}\,\frac{dm_e{}^2}{d\tau})]\eta + \frac{3}{\varepsilon^2}\,[\frac{d}{d\tau}\,m_e(\theta_0 + \frac{54m_e{}^2}{\varepsilon^6})] \\
&\quad [1 - \exp(-\frac{\varepsilon^2}{3m_e}\eta)]
\end{aligned} \qquad (32)$$

The unknown functions $\hat{M}_e$, $\hat{M}_S$, θ_S, ϕ_0 for $\tau \leq 0$ and m_e, m_S, θ_0, ϕ_0 for $\tau \geq 0$ are to be determined from the matching condition with the outer fields numerically calculated. Apparently, the latter are continuous from the former at $\tau = 0$, respectively.

The scales $\Delta\Psi$ and Δt still remain unknown. They are specified as follows. We note that the interfacial rate quantities, such as M_S, M_e, dY_S/dt, dY_{*S}/dt and dT_S/dt, must be independent of the choice of the scales ΔT, ΔY, $\Delta\Psi$ and Δt. Since dT_S/dt generally takes a non-zero finite value at the transition time, we choose them in such a way that $(dT_S/dt)/(T_c\lambda g/\rho gh T_g X_0{}^2)$ coincides with $d\theta_S/d\tau$. This leads to the relations $\Delta t = \Delta T$ and

$$\frac{\Delta\Psi}{\Delta Y} = \frac{\rho g\lambda g}{\rho_c\lambda_c} \frac{T_c h_{T_c}}{(-h_{Y_c})} (\frac{T_g - T_c}{T_c})^2 \tag{33}$$

which, along with Eqs.(19) and (20), express the scales $\Delta\Psi$ and Δt in terms of the parameter ε.

In view of Eqs.(22) and (28), Eq.(33) implies that M_S and M_e also take non-zero finite values at the transition time. In the present asymptotic analysis, the quantities $d\theta_S/d\tau$, $\hat{M}_S$ and $\hat{M}_e$ can be regarded as constants in the first approximation. Hence, in Eqs.(30) and (32), ϕ is found to be related with η linearly. It should be noted that the linear time change in θ_S brings about the singular behavior of $d\eta_{*S}/d\tau = -d\eta_S/d\tau$. They diverge at $\tau = 0$ in proportion to $(-\tau)^{-1/2}$.

The liquid-gas interafce in the binary component system is characterized by a jump in the concentration. The discontinuities existing in other properties seem to be secondary, in contrast with the case of one-component system in which the interface is characterized by the density jump. The diverging concentration gradient at the critical interface, therefore, may be considered as a remainder of this concentration discontinuity. In other words, the transition region is a simple continuation of the liquid-gas interface the thickness of which grows from "zero" to $O(\Delta T)$. This is evident from the fact that the interfacial conditions (27) and (28) are identical, in form, to the once integrated transport equations (25) and (26) in the transition region. In the subcritical evaporation regime the discontinuity is maintained by the action of thermodynamic forces involved in the phase equilibrium relations. Once this mechanism disappears at the transition, the diverging concentration gradient tends to induce a large diffusion flux but the vanishingly small diffusion coefficient can keep the gradient stationary. The quasi-steady form of Eq.(25) may be interpreted in this way. Since the unsteady term in the mass transport equation is smaller than the

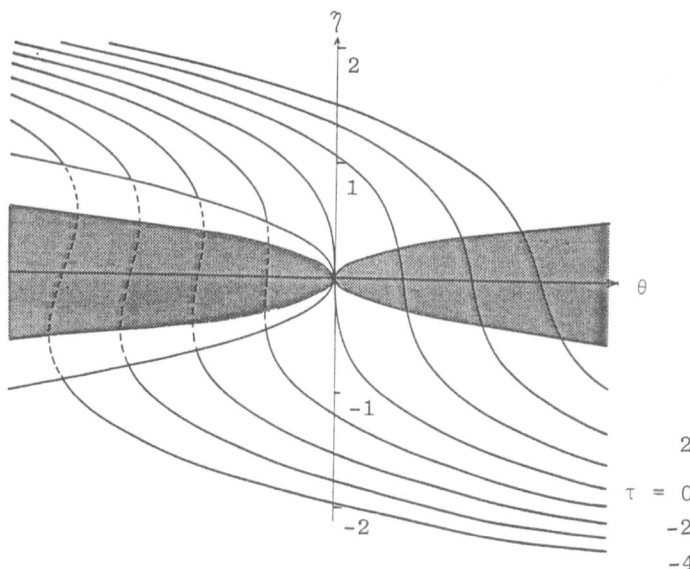

Figure 4

diffusion term by $O(\Delta T)$, it has no significant influence on the mass
transport within the thin transition region even though it plays an
important role in the outer fields. On the other hand, the situation
is a little bit different for the heat transport. Since the heat of
mixing, h_Y, takes a non-zero finite value at the critical point, the
heat produced by the concentration change must be compensated by the
conductive and convective heat fluxes.

   Although the concentration profile $Y(t,T)$ is defined outside of the
saturation curve, it will be interesting to continue the profile into
the co-existing region. This can be done by taking into account the
higher-order terms neglected in the solution (29). In this case, the
solution assumes the same form as Eq.(31) and can be approximated by

$$\theta = \theta_S - \frac{\varepsilon^2 \theta_S}{3\hat{M}_e} \eta - \frac{1}{3\hat{M}_e} \eta^3 \tag{34}$$

near the critical point. Figure 4 shows its graph for $d\theta_S/d\tau(\tau=0) =$
$M_e = 1$. A considerably large value of ε has been taken to emphasize the
behavior in the co-existing region. The figure should be compared with
the isothermal lines in the PV-diagram of a pure substance. We see
that the temperature, pressure and density in the PV diagram,
respectively, corresponds to the time, temperature and concentration
in Fig.4.

In Fig.4 the boundaries of the shadowed regions are given by the
equation $\theta = \pm 3\eta^2/\varepsilon^2$, where the temperature change effect on the
diffusion coefficient becomes as significant as the concentration change
effect. In the subcritical evaporation regime the boundary coincides
with the line which is formed by connecting the extreme points of the
concentration profiles. According to the analogy to the pure substance
case, it may be considered to give the spinoidal curves for the binary
component system. If so, unlike the pure substance case the degree
of superheat or supercool is found to vary with the condition of
realization because it depends on the value of $\hat{M}_e$. In the supercritical
evaporation regime the concentration gradient $\partial\eta/\partial\theta$ on the θ-axis
decreases with inctreaing temperature. This attributes to the dependence
of the diffusion coeficient on temperature.

Concluding Remarks

The physical mechanism of the transition from sub- to supercritical
evaporation regime is summerized as follows. The non-zero finite heat
flux near the surface rises the surface temperature through the critical
temperature. This heating process is almost the same as the pure heat
transport process and has no significant influence from the diffusion
process. The vanishing diffusivity near the critical point can bridge,
in a continuous way, the subcritical discontinuous liquid-gas interface
and the supercritical spatially continuous phase change. At this
unsteady transition the heat of mixing also plays an important role in
the detailed balance of energy.
    The analysis shows that the heat and mass transfer rates become
almost of the same order near the critical interface. This is especially
so for the case of Stage II which have been omitted in the present
paper. When we take the local thermal and mass diffusivities to estimate
the magnitudes of the heat and mass transfer rates, there appears a
great difference between their characteristic time scales. This is
because the thermal diffusivity, $\alpha = \lambda/\rho h_T$, has a non-zero value at
the critical point whereas the mass diffusivity D vanishes. If the
gradients of temperature and concentration in the physical space are
comparable, the predicted heat and mass transfer rates will differ
greatly. The concentration gradient $\partial Y/\partial x$ is, however, greater than
the temperature gradient $\partial T/\partial x$ by the factor $\partial Y/\partial T$ of $O(\Delta Y/\Delta T)$, so
that this factor makes the mass transfer rate comparable with the heat
transfer rate near the critical interface. Using the modified specific

heat defined by Eq.(5), we can make another interpretation for this. The energy equation may be expressed in terms of the modified specific heat c_p in the form $\rho_c c_p DT/Dt = \partial/\partial x(\lambda \partial T/\partial x)$. Near the critical interface c_p becomes great due to the divergence nature of DY/DT, so that the apparent thermal diffusivity $\lambda/\rho c_p$ becomes the same order as the mass diffusivity. In fact, it follows Eq.(11) and (12) that the modified Lewis number $L_e = \lambda/\rho c_p D$ must become equal to unity at the critical liquid-gas interface. A merit of using the modified specific heat, which is naturally introduced in the present formalism based on the state-space description, lies at this point. It is also useful for numerical calculation. In numerical calculation it is important that there are no quite different time scales in the same problem. Since the time scale of the calculation is usually based on the thermal diffusivity of the undisturbed gas, such a calculation based on the conventional formalism will encounter a sever difficulty near the critical condition.

Acknowledgement

This study was performed during the author's stay in West Germany as a Alexander von Humboldt research fellow. The author thanks Profs. F.H.Busse and H.Schamel for their discussion.

References

1. T.Kadota & H.Hiroyasu, Trans. Japan Soc. Mech. Engrs.,44,3885 (1978)
2. D.B.Spalding, ARS J., 29, 825 (1959)
3. P.R.Wieber, AIAA J., 1, 2764 (1963)
4. D.E.Rosner, AIAA J., 5, 163 (1967)
5. J.A.Manrique & G.L.Borman, Int. J. Heat and Mass Transf., 12, 1081 (1969)
6. G.M.Feach et al., 12th Symp. (Int.) on Com., (1969), 9
7. R.L.Matiosz, S.Leipziger & T.P.Torda, Int. J. Heat and Mass Transf., 15, 831 (1972)
8. H.Hiroyasu et al., Trans. Japan Soc. Mech. Engrs., 40, 3147 (1974)
9. T.Kadota & H.Hiriyasu, Trans. Japan Soc. Mech. Engrs.,46,1591 (1980)
10. T.A.Brzustowski, Canad. J. Chem. Engng., 43, 30 (1965)
11. F.A.Williams, Combustion Theory, Addison-Wesley (1965)
12. A.Umemura, 21st Symp.(Int.) on Combustion,The Comb. Inst. (1987)
13. S.R.De Groot & P.Mazur, Non-Equilibrium Thermodynamics, North Holland (1962)

HIGH TEMPERATURE EXTINCTION OF
PREMIXED FLAMES

David W. Mikolaitis
Department of Engineering Sciences
University of Florida
Gainesville, FL   32611

ABSTRACT

The far field of a stretched premixed flame with $A \rightarrow B \rightleftharpoons C$ kinetics is studied. Each of the reaction rates are assumed to be proportional to the mass fraction of the reactants and have a temperature dependence that is continuous, positive, and vanishes at both infinite and zero absolute temperature. Arrhenius reaction rate functions are examples that satisfy these requirements. Bounds are uncovered for the amount of blowing from behind the flame that plane premixed flames can withstand for all Lewis numbers when the intermediate species are lighter than the deficient reactant. When the intermediate products are heavier than the deficient reactant, bounds on the blowing are found when the Lewis number of the deficient reactant is less than one.

INTRODUCTION

Through the analysis of the far field of adiabatic stretched premixed flames with $A \rightarrow B$ kinetics, we have been able to show that premixed flames with Lewis numbers less than one will be extinguished given sufficiently strong blowing from behind [1]. This is inferred from the analysis in that the mass fraction of combustible becomes negative in the far field if the opposed flow is above a maximum allowable level. The activation energy asymptotic (AEA) analyses of this case [2-7] do not show true extinction. What we mean by true extinction is the failure of existence for a flame structure for sufficiently strong opposed flow.

In the previous work [1] it was unclear as to what effect dissociation would have on the results. Here we will investigate a kinetic mechanism that incorporates dissociation. The simplest possible reaction mechanism that would include dissociation is $A \rightleftharpoons B$.

It seems unlikely that the analysis of such a mechanism would reveal anything of interest in that the final equilibrium state would be composed of major species only without any intermediates. A more plausible model is the A→B→C mechanism where B represents intermediate products and the final equilibrium state is dominated by intermediate and final products.

The plan of this paper is to first analyze the far field of the stretched A→B→C flame so that the case without dissociation is fully documented. Once the non-dissociating model is fully explored we shall include the effects of dissociation.

MATHEMATICAL MODEL

The equations that model two-dimensional, low Mach number combustion waves under the assumption of calorifically and thermally perfect gases with constant transport properties are

$$\partial(\rho u)/\partial x + \partial(\rho v)/\partial y = 0 \tag{1}$$

$$\partial(\rho uT)/\partial x + \partial(\rho vT)/\partial y = \nabla^2 T + q_1 Yf_1(T) + q_2 Xf_2(T) - q_2 Zf_3(T) \tag{2}$$

$$\partial(\rho uY)/\partial x + \partial(\rho vY)/\partial y = (L_A)^{-1}\nabla^2 Y - Yf_1(T) \tag{3}$$

$$\partial(\rho uX)/\partial x + \partial(\rho vX)/\partial y = (L_B)^{-1}\nabla^2 X + Yf_1(T) - Xf_2(T) + Zf_3(T) \tag{4}$$

$$Z = 1 - X - Y \tag{5}$$

$$\rho u\partial u/\partial x + \rho v\partial u/\partial y = -\partial p/\partial x + P(\nabla^2 u + (\partial(\partial u/\partial x + \partial v/\partial y)/\partial x)/3) \tag{6}$$

$$\rho u\partial v/\partial x + \rho v\partial v/\partial y = -\partial p/\partial y + P(\nabla^2 v + (\partial(\partial u/\partial x + \partial v/\partial y)/\partial y)/3) \tag{7}$$

$$\rho = 1/T. \tag{8}$$

u and v are the non-dimensionalized x and y components of fluid velocity, T is the non-dimensional temperature, ρ is the non-dimesional density, p is the non-dimensional pressure, P is the Prandtl number and X, Y and Z are the mass fractions of the intermediate product, the reactant and the final product respectively. These equations are continuity (1), energy balance (2), species balance (3-5), x and y momentum balance (6-7) and equation of state (8).

Proceeding as in [1], this system can be simplified for plane flames in a non-uniform flow field. The resulting equations under the limit of large heat release are

$$MdT/dx = d^2T/dx^2 + q_1 Yf_1(T) + q_2 Xf_2(T) - q_2(1-X-Y)f_3(T) \tag{9}$$

$$MdY/dx = (1/L_A)d^2Y/dx^2 - Yf_1(T) \tag{10}$$

$$MdX/dx = (1/L_B)d^2X/dx^2 + Yf_1(T) - Xf_2(T) + (1-X-Y)f_3(T) \tag{11}$$

$$Pd^2(TdM/dx)/dx^2 - Md(TdM/dx)/dx + T(dM/dx)^2 = 0. \tag{12}$$

M is the mass flux, ρu. Without the limit of large heat release

$(T_{flame}/T_{remote\ reactants} \gg 1)$ the right hand side of equation (12) could be a non-zero constant.

Appropriate boundary conditions are

$$T(-\infty)=X(-\infty)=0, \quad Y(-\infty)=Y_{-\infty}, \quad T(+\infty)=T_f, \quad M(+\infty)=M_o$$

$$X(+\infty)=f_3(T_f)/(f_2(T_f)+f_3(T_f)), \quad Y(+\infty)=0 \tag{13}$$

where T_f is the temperature of the remote products, a quantity that must be found as part of the analysis. In addition, a boundary condition is needed to force the balance in the remote reactants to be one between conduction and convection without appreciable reaction. Such a condition is

$$(d^2T/dx^2)/(M\ dT/dx) \to 1 \text{ as } x \to -\infty. \tag{14}$$

THE REMOTE BURNED STATE

Since the system (9-13) is autonomous, we can transform the problem into a phase space problem on a finite domain with independent variable T through the definition

$$z = dT/dx. \tag{15}$$

The resulting system of equations is

$$M = dz/dT + (q_1Yf_1(T) + q_2Xf_2(T) - q_2(1-X-Y)f_3(T))/z \tag{16}$$

$$MdY/dT = (1/L_A)\ d(zdY/dT)/dT - Yf_1(T)/z \tag{17}$$

$$MdX/dT = (1/L_B)\ d(zdX/dT)/dT +$$
$$(Yf_1(T) - Xf_2(T) + (1-X-Y)f_3(T))/z \tag{18}$$

$$Pzd(zd(TzdM/dT)/dT)/dT - Mzd(TzdM/dT)/dT + Tz^2(dM/dT)^2 = 0 \tag{19}$$

and the boundary conditions at $x=+\infty$ become

$$z(T_f)=0, \quad X(T_f)=f_3(T_f)/(f_2(T_f)+f_3(T_f))=X_f,$$

$$Y(T_f)=0, \quad M(T_f)=M_o. \tag{20}$$

Near $T=T_f$ we can write

$$z = \epsilon\varsigma, \quad Y = \epsilon\psi, \quad X = X_f + \epsilon\xi$$

$$T = T_f + \epsilon\tau, \quad M = M_o + \delta\mu. \tag{21}$$

The resulting system is

$$M_o = d\varsigma/d\tau + q_1\psi f_1(T_f) +$$
$$q_2[\tau\Lambda(T_f) + \psi f_3(T_f) + \xi(f_2(T_f)+f_3(T_f))] \tag{22}$$

$$M_o\ \varsigma\ d\psi/d\tau = (1/L_A)\ \varsigma\ d(\varsigma\ d\psi/d\tau)/d\tau - \psi f_1(T_f) \tag{23}$$

$$M_o\ \varsigma\ d\xi/d\tau = (1/L_B)\ \varsigma\ d(\varsigma\ d\xi/d\tau)/d\tau + \psi f_1(T_f) -$$
$$\xi(f_2(T_f)+f_3(T_f)) - \psi f3(T_f) - \tau\Lambda(T_f) \tag{24}$$

and another equation for μ that is of no consequence in the remainder of the analysis. The function Λ is defined by

$$\Lambda(T) = (f_3\ df_2/dT - f_2\ df_3/dT)/(f_2 + f_3). \tag{25}$$

The boundary conditions are

$$\varsigma(0)=\psi(0)=\xi(0)=0. \tag{26}$$

The singular point described by the system (22-24, 26) is inherently non-linear, but fortunately it admits solutions of the form

$$\psi = a\tau, \quad \xi = b\tau, \quad \zeta = c\tau. \tag{27}$$

Substitution into (22-24) results in

$$c = [L_A M_0 - \sqrt{(L_A M_0)^2 + 4L_A f_1(T_f)}]/2 \tag{28}$$

and the set of algebraic equations

$$a(q_1 f_1(T_f) + q_2 f_3(T_f)) + bq_2(f_2(T_f) + f_3(T_f))$$
$$= M_0 c - c^2 - q_2 \Lambda(T_f) \tag{29}$$

$$a(f_1(T_f) - f_3(T_f)) + b(c^2/L_B - M_0 c - f_2(T_f) - f_3(T_f)) = \Lambda(T_f) \tag{30}$$

that are trivially solved for a and b so that a,b and c are now known as functions of M_0 and T_f. There is another solution for c with the positive root in (28), but that solution is not consistent with the fact that T approaches a constant as x goes to infinity.

THE REMOTE SINGULAR POINT FOR NO DISSOCIATION

The solution fails to be physically realistic when a is positive since that would correspond to a negative mass fraction for Y near the singular point. The case where a=0 therefore defines a critical condition.

The specification of the critical condition is very simple in the case of no dissociation. When dissociation is neglected we must take $f_3(T) = 0$ which implies that $\Lambda = 0$. Setting a equal to zero and eliminating b from (29-30) gives rise to

$$(M_0 c(1-L_A) - f_1(T_f) L_A)(M_0 c(L_A/L_B - 1) + L_A/L_B \ f_1(T_f) - f_2(T_f)) = 0. \tag{31}$$

Substituting (28) for c gives the two critical conditions

$$M_0{}^2{}_c = L_A f_1(T_f)/(1-L_a) \tag{32}$$

and

$$M_0{}^2{}_c = (f_2(T_f) - L_A/L_B \ f_1(T_f))^2/[L_A(f_2(T_f) - f_1(T_f))(L_A/L_B - 1)]. \tag{33}$$

Since the Lewis numbers and reaction rates are always positive, condition (32) can only be attained when $L_A < 1$. In addition, if we only consider the case where $f_2(T) > f_1(T)$ throughout the range of possible final product temperatures T_f so that product formation is a faster reaction than the breakdown of the reactants (usually a good assumption since the breakdown of the relatively stable reactants is governed by a fairly high activation temperature mechanism and the formation of product from intermediates is typically through reactions with small activation temperatures), then condition (33) can only apply if $L_A > L_B$. This is a very common occurence since it requires, roughly speaking, the mean molecular weight of the intermediate species to be smaller than the molecular weight of the deficient

reactant. The only potential mechanism that we can think of that does
not conform to this condition is lean hydrogen oxidation where the
deficient reactant has molecular weight 2 and the intermediate species
are H, OH and O. It is likely that the mean molecular weight of the
intermediates will be greater than 2 in this case.

Condition (32) is identical to the critical condition for
the A→B mechanism [1].

An example where $L_B < L_A$ and $L_A > 1$ is given in Figure 1.
This situation occurs when the intermediate products are ligther than
the deficient reactant and the deficient reactant is heavier than the
bulk gas. In this case only (33) defines a limiting condition. In
this and all subsequent figures, the shaded regions are regions in the
parameter plane where the remote reactant mass flux is mathematically
negative and hence is of no physical significance.

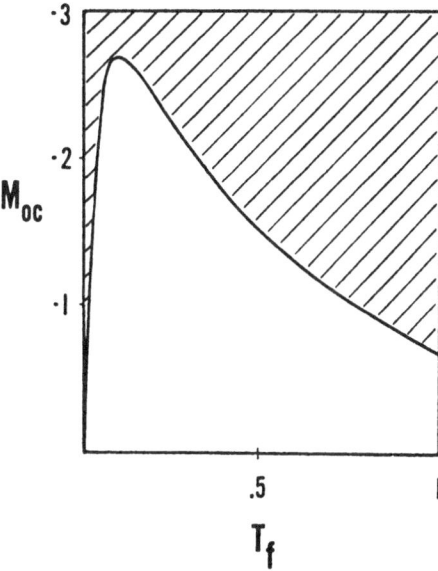

Fig. 1: Critical remote mass flux versus flame temperature.
$L_A=1.2$, $L_B=.8$, $q_1=-1$, $q_2=2$, $f_1=\exp(-1/T)/T$, $f_2=\exp(-.1/T)/T$

Figure 2 shows a case when $L_A < 1$ and $L_A < L_B$. Since this
case only applies to the situation where the deficient reactant is
both lighter than the intermediate products and the bulk gas, it seems
likely that it may only apply to lean hydrogen/oxygen or lean hydro-
gen/oxygen/diluent mixtures. In this case only the condition (32)
applies.

Figure 3 shows a case where $L_B < L_A < 1$. Such a situation
is likely for lean methane/air flames, for example, where the prin-

ciple intermediates are lighter than the deficient reactant which in turn is lighter than the bulk fluid. In this case critical conditions are given by both (32) and (33).

In the first two cases it is clear that there are no physically realizable solutions if the remote mass flux in the burned gas becomes too high. In the last case shown in Figure 3, the mass fraction of the remote reactants is positive if conditions are such that the

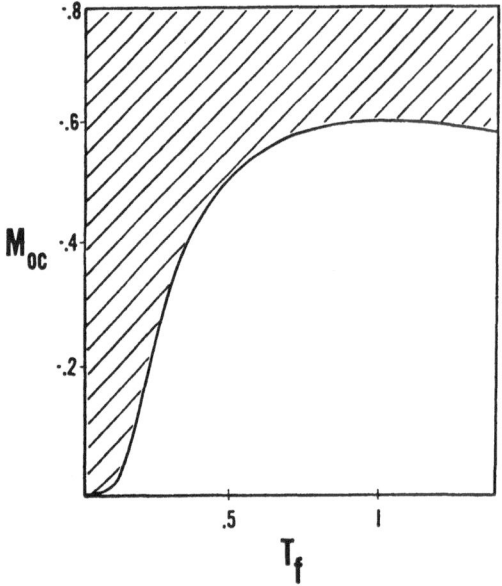

Fig. 2: Critical remote mass flux versus flame temperature.
L_A=.5, L_B=.7, q_1=-1, q_2=2, f_1=exp(-1/T)/T, f_2=exp(-.1/T)/T

flame is above both curves. As a practical matter, however, such solutions are probably not accessible since the starting conditions in any real system would be in the lower domain and there is no way to enter the upper domain in any continuous manner except through the single point where the two domains meet. The "solution" at the point where the two solutions meet is found as limit of problems where the remote upstream reactant mass fraction goes to zero and hence is not truly a flame. In fact, the solid curves are not part of the region of physically realistic flame solutions because they correspond to situations where the remote upstream mass fraction of reactant is zero and hence there is no flame. This fact is found through numerical integration of the governing equations.

THE REMOTE SINGULAR POINT WITH DISSOCIATION

When dissociation is retained, the critical M_O is found by solving the algebraic equation

$$(CM_O(1-L_A)-L_Af1(Tf)-q_2\wedge(T_f))(CM_O(-1+L_A/L_B)+f_1(T_f)L^A/L_B$$
$$-f_2(T_f)-f_3(T_f)) = q_2\wedge(T_f)(f_2(T_f)+f_3(T_f))(34)$$

where C, a function of M_O, is given by (28). This is done most easily by numerical root finding.

First we will consider the case where $L_B < L_A$ and $L_A > 1$, such as a lean heavy hydrocarbon/air premixed flame, with weak dissociation.

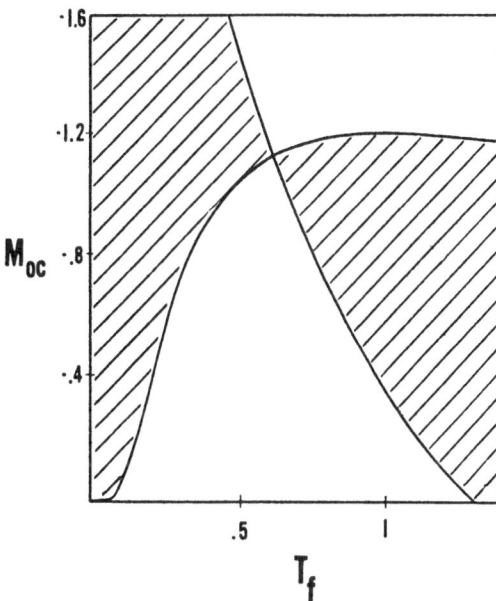

Figure 3: Critical remote mass flux versus flame temperature.
$L_A=.8$, $L_B=.4$, $q_1=-1$, $q_2=2$, $f_1=\exp(-1/T)/T$, $f_2=\exp(-.1/T)/T$

A typical parameter plot is shown in Figure 4. Comparing this plot to the corresponding case without dissociation (Figure 1) we see that dissociation decreases the maximum amount of blowing that the flame can withstand.

In the case where $L_A < L_B < 1$, as shown in Figure 5, the maximum possible amount of blowing is increased through the action of dissociation. As mentioned earlier, such an ordering of the Lewis numbers is very infrequent in applications. One of the only possible physical systems that might have such an ordering is lean hydrogen/air premixed flames.

The case where $L_B < L_A < 1$ is shown in Figure 6. This manner of

Lewis number ordering is characteristic of rich propane or heavier
hydrocarbon/air mixtures. It is interesting to note that there is no
longer a maximum amount of blowing predicted by the analysis of the

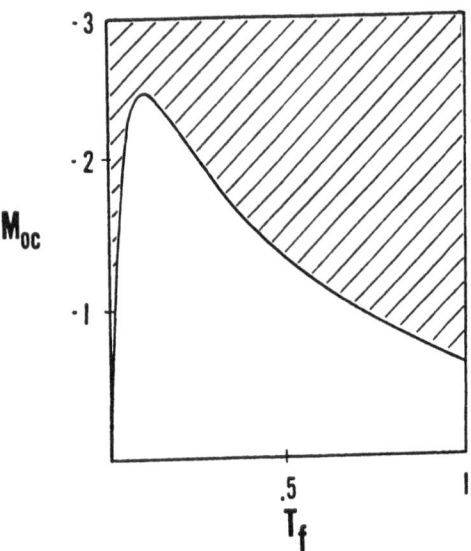

Figure 4: Critical mass flux versus flame temperature
$L_A=1.2$, $L_B=.8$, $q_1=-1$, $q_2=2$, $f_1=\exp(-1/T)/T$
$f_2=\exp(-.1/T)/T$, $f_3=.001\ \exp(-2/T)/T$

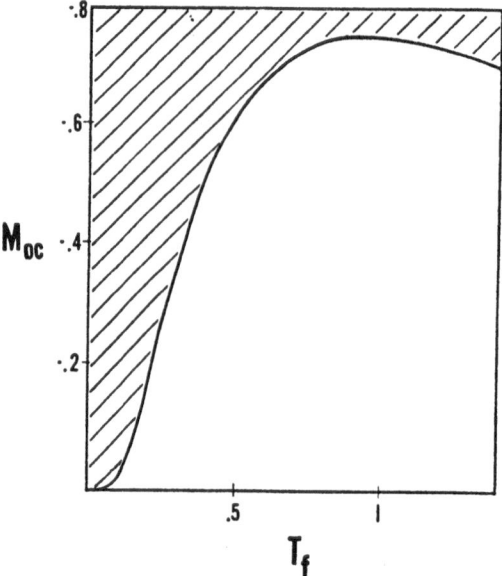

Figure 5: Critical mass flux versus flame temperature.
$L_A=.5$, $L_B=.7$, $q_1=-1$, $q_2=2$, $f_1=\exp(-1/T)/T$
$f_2=\exp(-.1/T)/T$, $f_3=\exp(-2/T)/T$

remote burned state. A gap opens up through which solution curves
may pass and the stronger the rate of dissociation, the wider the gap.
It is probable,though we hasten to add that we have not checked this
result fully, that the situation can arise where near stoichiometric
mixtures will extinguish with sufficiently high blowing but rich
mixtures will not with some pivotal value of the remote mass fraction
of reactant, denoted by Y_C, that seperates the two cases. This
phenomena is shown schematically in Figure 7. As the rate of dissociation is increased, the gap should widen with a subsequent increase
in the pivotal value of the remote reactant mass flux. With a
sufficiently high rate of dissociation, it seems likely that the
pivotal value of the mass fraction of reactants should go to 1 with
further increases in the rate of dissociation resulting in no maximum
rate of blowing.

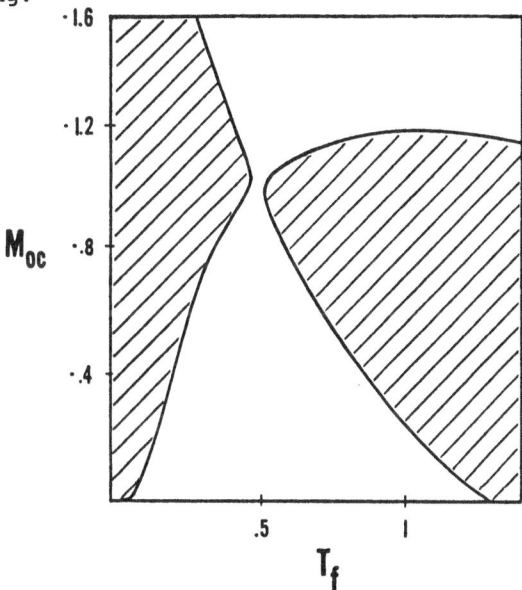

Figure 6: Critical mass flux versus flame temperature.
$L_A=.8$, $L_B=.4$, $q_1=-1$, $q_2=2$, $f_1=\exp(-1/T)/T$
$f_2=\exp(-.1/T)/T$, $f_3=.001 \exp(-2/T)/T$

CONCLUDING REMARKS

Here we have analyzed the response of a plane premixed flame with
sequential kinetics to a non-uniform flow field. Two seperate cases
were investigated; one where the second step was irreversible and one
where it was reversible. The inclusion of reversibilty in the second
step did not alter the qualitative flame response except in the case

where the Lewis number ordering was $L_B < L_A < 1$ and the mass fraction of reactants in the remote supply stream was below some pivotal value. Physically this case corresponds to sufficiently lean methane/air or sufficiently rich heavy hydrocarbon/air mixtures. The mathematical result suggests that there is a well defined equivalence ratio for methane/air mixtures below which flames can withstand large amounts of blowing but above which they cannot. The pivotal equivalence ratio should be less than one, i.e. lean, in this case. Similarly, there should exist a well defined equivalence ratio for each heavy hydrocarbon/air mixture above which flames can withstand large amounts of blowing but below which they cannot. In this case the pivotal equivalence ratio should be on the rich side of stoichiometry.

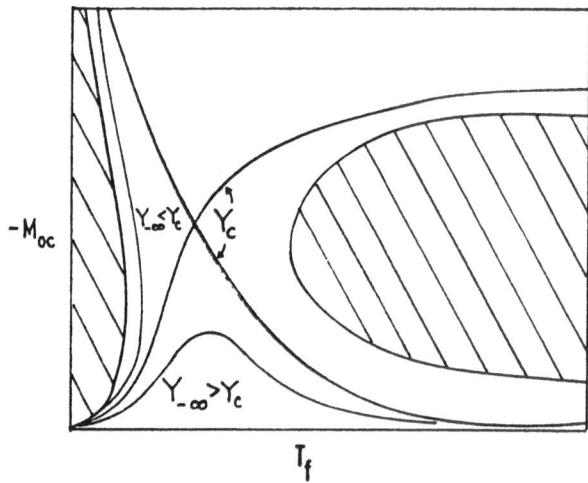

Figure 7: Schematic of flame response curves for the case where $L_B < L_A < 1$.

ACKNOWLEDGEMENT

We would like to acknowledge that this work was supported by the Air Force Office of Scientific Research, Air Force Systems Command, USAF, under grant number AFOSR 87-0236. The US Government is authorized to reproduce and distribute reprints for Governmental purposes notwithstanding any copyright notice thereon.

REFERENCES

[1] Mikolaitis, D. (1987). On the abrupt extinction of premixed flames with Lewis numbers less than one. (Submitted for publication.)

[2] Buckmaster, J. (1979). The quenching of a deflagration wave held in front of a bluff body. Seventeenth Symposium (International) on Combustion, The Combustion Institute, p. 835.

[3] Buckmaster, J. and Mikolaitis, D. (1982). The premixed flame in a counterflow. Combust. Flame, 47, 191.

[4] Libby, P., Liñán, A., and Williams, F.A. (1983). Strained premixed laminar flames with non-unity Lewis numbers. Combust. Sci. Tech., 34, 49.

[5] Libby, P. and Williams, F.A. (1982). Structure of laminar flamelets in premixed turbulent flames. Combust. Flame, 44, 287.

[6] Libby, P. and Williams, F.A. (1983). Strained premixed flames under nonadiabatic conditions. Combust. Sci. Tech., 31, 1.

[7] Libby, P. and Williams, F.A. (1984). Strained premixed flames with two reaction zones. Combust. Sci. Tech., 37, 221.

A MODEL FOR LOWER DEFLAGRATION LIMIT AND BURNING RATE
OF CATALYZED AMMONIUM PERCHROLATE

Tohru Mitani

National Aerospace Laboratory, Kakuda Branch

Ohgawara, P.O. #7, Miyagi 989-12, Japan

ABSTRACT

An analytical model has been developed to describe the burning behavior of catalyzed ammonium perchlorate(AP) for pressures below 10 MPa. The increase in both the burning rate and the lower deflagration limit pressure(LDP) of AP with catalyst content is investigated. The peculiar increase of LDP of AP by catalysts càn be understood on the basis of an inherent instability of the condensed phase deflagration wave decoupled from the gas phase deflagration wave. The effects of external heat addition, preheating and the fuel added to AP on the LDP are also explained using the model.

1. INTRODUCTION

Ammonium perchlorate(AP, NH_4ClO_4) is the primary oxidizer in composite propellants. The self-deflagration rate of AP at typical pressures is approximately the same as the burning rate of many AP-based propellants. Therefore, it is believed that AP deflagration may well be a controlling factor for the rates of propellants. AP undergoes self-sustained combustion only in certain pressure ranges above the lower deflagration limit pressure(LDP). An average value of 2 MPa has been measured for the LDP at room temperature.

There have been some pivotal studies explaining the LDP of AP and propellants(1,2,3). These models predict a LDP based on heat loss, but agreement with the experimental value was obtained by assuming an unjustifiably large heat loss in the solid. Hightower and Price reported the existence of thin decomposing melt on the surface of AP under normal self-deflagration condition(4). They indicated that the LDP corresponds to attainment of heating rates so low that AP decomposes without reaching a melting temperature. Guirao and Williams developed an analytical model in which exothermic condensed phase reactions occurring in the liquid layer are responsible for the deflagration of AP at pressures from 2 to 10 MPa(5).

It should be remembered that the LDP of AP is very sensitive to cata-lysts. Paradoxically, most catalysts which are used to increase burning rates of AP-composite propellants _increase_ the LDP of AP. A trace of any such catalyst makes AP non-ignitable at a pressure of 4 MPa(a typical operating pressure of solid propellant rockets). Friedman et al. examined many catalysts, in which the addition of 0.3 wt % of copper chromite in-ceases the LDP from 2 MPa to 17 MPa(6). Shadman-Yazdi and Petersen inves-tigated the effect of catalysts on the LDP(7). Boggs et al. used AP doped with K^+, $Cr_2O_7^{2-}$ and MnO_4^- in their studies(8). Their cinephotographic and electron microscopic observations show the importance of physical proc-esses such as melt formation at the burning surface.

The exact mechanism, or even the rate determining processes, of AP def-lagration are not clear, even today. Consequently, the sites at which burning rate catalysts act to enhance propellant burning rates are the topic of much current discussion. In this paper, a deflagration model for monopropellants is generalized for catalyzed AP to explain the pe-culiar effects of catalysts.

2. MODELS FOR LDP

The critical regression rate of AP at the LDP is found to be about 3 mm/sec, which is an order of magnitude higher than the critical rates of other monopropellants(9). Since heat loss rate in the deflagration wave is related to residence time in the wave, the contribution of heat loss to the deflagration is inversely proportional to the square of the propagation rates of the waves. This implies that the effect of heat loss rapidly diminishes as the regression rates increase. Unusually low heat loss at the LDP was found in experiments on quenching diame-ters of AP and HAP(Hydroxyl Ammonium Perchlorate; $HONH_3ClO_4$)(9). These results suggest that the increase of radiative heat loss might be a consequence from drastic changes in burning behavior of AP but can not be the cause of the LDP at least with AP and HAP.

The burning rates of AP composite propellants can be augmented by catalysts at 4 MPa(below the LDP of catalyzed AP). In the burning of AP composite propellants, the diffusion flame formed between AP and the fuel binder is not directly affected by catalysts. The heat feedback from the diffusion flame enhances the decomposition flame and thus stabilizes the gas phase deflagration wave. This suggests that the inherent ability of catalysts to increase reaction rates may be revealed only when the deflagration wave is stabilized.

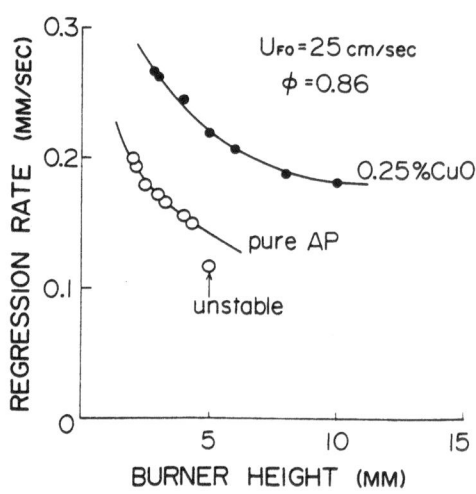

Figure 1: Augmented burning rate of catalyzed AP measured by a counterflow AP burner at atmospheric pressure

Catalysis of CuO in AP can be seen below the LDP of AP catalyzed by CuO as shown in Fig. 1. Figure 1 shows burning rate augmentation resulting from addition of CuO observed using a counterflow AP burner at atmospheric pressure(10). The AP decomposition flame in the gas phase, located 300 μm above the burning surface, was stabilized by the addition of heat from the premixed burner above an AP pellet. This experiment shows that the regression rate of AP is increased by 50 % with the addition of 0.25 wt.% CuO. Augmented burning rate induced by catalysts below the LDP were observed using AP samples with a counterflow diffusion flame(11) and AP samples with a minor addition of fuels (12). Thus the catalysts inherently increase reaction rates even below the LDP as well as above the LDP. A question remains, however, as to why the reaction rates accelerated by catalysts inhibits the deflagration of AP in lower pressure and causes the irregular increase of LDP of AP.

This contradictory phenomenon must be understood in light of the stability of deflagration waves. AP is a monopropellant and the burning is self-supported by two essential exothermic reactions, one in the condensed and the other in the gas phase(5, 13). These two reaction waves are strongly coupled in the steady deflagration, and the decoupling may cause an inherent instability of the condensed phase deflagration waves. Because molecular diffusion in the condensed phase is negligible, an excess enthalpy accumulated ahead of the reaction zone in the condensed phase destabilizes the condensed phase deflagration waves, if the heat addition from the gas phase to the surface decreases. This intrinsic instability has been studied theoretically(14, 15), and spinning and pulsating propagations have been observed experimentally for gasless combustion in the Ni-B system, for example(16).

3. EFFECTS OF CATALYSTS ON AP BURNING

A model involving exothermic decomposition in the condensed phase, equilibrium vaporization and exothermic combustion in the gas phase has been developed for describing the flame structure of nitramines(17). Because there are common characteristics between nitramines and AP, e.g. a melting layer, strong exothermicity in the condensed phase and gaseous flame, this formulation and results of the model can be applied to the AP deflagration.

An overall energy balance for the condensed phase can be written as

$$\lambda_g (dT/dx)_s = \dot{m} \left(\int_{T0}^{Ts} c dT - q_s \right) , \qquad (1)$$

where the effective heat in the condensed phase(q_s) is expressed by

$$q_s = q_c (1 - G) - l_v G . \qquad (2)$$

A fraction of AP must vaporize at the burning surface and be consumed in the gaseous decomposition flame. The fraction of AP reacting in the gas flame is expressed by G. The effective heat (q_s) can be experimentally evaluated from temperature profiles and might be positive(exothermic) or negative(endothermic) depending on q_c, l_v and G. The properties q_c and l_v are determined by thermodynamics and they are related to the heat release by gas phase reaction(q_g) by

$$q_g = q_c + l_v \quad . \tag{3}$$

The heat of reaction in the condensed phase is nondimensionalized using the enthalpy change in the condensed phase and expressed as

$$\alpha = q_c / \int_{T_0}^{T_s} c \ dT \quad . \tag{4}$$

Since the temperature at the burning surface is lower than that of final gaseous products, the value of α must greater than unity.

The regression rate of AP due to the condensed phase decomposition is found in Ref. 17 and expressed using α and G as

$$r \propto [\ \omega_c / (\ \alpha (1-G) + (1-\alpha) \ln (1/G))]^{1/2} \quad . \tag{5}$$

The modification of regression rate by the addition of catalysts is found to be

$$\kappa \equiv \frac{r}{r_0} = \hat{\omega}_c^{1/2} [\frac{\alpha (1-G_0) + (1-\alpha) \ln (1/G_0)}{\alpha (1- G) + (1-\alpha) \ln (1/ G)}]^{1/2} \quad . \tag{6}$$

The subscript(0) indicates the properties under catalyst-free conditions. The reaction rate and the nondimensional heat of reaction are dependent on the surface temperature. No change in T_s has been assumed in Eq. 6, and the effect of catalysts on ω_c is represented by the modification of the preexponential factor in the reaction rate law. The increase of reaction rate in the condensed phase caused by the catalysts is denoted by $\hat{\omega}_c = \omega_c / \omega_{c0}$.

Using an overall gaseous reaction rate(ω_g), the consumption rate of gaseous AP can be expressed as

$$r \propto \omega_g^{1/2} /G \quad . \tag{7}$$

The augmentation in regression rate in the gas phase caused by catalysts is given by

$$\kappa = (\ G_0 \ / \ G \ ) \ \hat{\omega}_g^{1/2} \quad , \tag{8}$$

where the augmentation in the gas phase reaction rate is defined by $\hat{\omega}_g = \omega_g / \omega_{g0}$.

The steady state condition between the condensed and the gas phases yields the final results on the regression rate augmentation by catalysts, which is

$$\hat{R} \left[\frac{\alpha(1-G_0/\hat{R})+(1-\alpha)\ln(\hat{R}/G_0)}{\alpha(1-G_0)+(1-\alpha)\ln(1/G_0)} \right]^{1/2} = (\frac{\hat{\omega}_c}{\hat{\omega}_g})^{1/2} \quad , \quad (9)$$

where $\hat{\kappa} = \kappa / \hat{\omega}_g^{1/2}$ and a parameter, G, has been eliminated using Eqs. 6 and 8. In Eq. 9, the parameter denotes the contribution of the gas phase reaction to the modification on regression rate of AP and the other new parameter ($\hat{\omega}_c / \hat{\omega}_g$) is the relative sensitivity of the condensed phase reaction and the gas phase reaction to catalysis.

As seen from Eq. 5, there is no solution for extremely small values of G because $(1 - \alpha)$ is always negative. However, stable deflagration becomes impossible before G approaches the limit. The adiabatic situation behind the condensed phase deflagration wave is achieved when the following condition is satisfied:

$$\alpha (1 - G) = 1 \quad . \quad (10)$$

Even under this condition, the heat addition from the gas may exist(see Eq. 1). However, the heat supplied is completely consumed to vaporize the unreacted AP and it can not participate in the reaction in the condensed phase. Thus, the deflagration wave in the condensed phase is decoupled from the gas phase and becomes unstable(14 and 15). This inherent instability might be planar but pulsative or laterally propagating (spinning). These unstable burning modes are actually observed in the experiments using the counterflow AP burner. The pulsative, galloping and spinning instabilities generally appear near extinction limits not only in the AP deflagration but also in premixed flames, diffusion flames and detonations. If the stability limit can be represented by the adiabatic condition behind the condensed phase reaction wave, the stable burning region of catalyzed AP is derived from Eqs. 8 and 10 as follows:

$$\hat{\kappa} < G_0/(1 - 1/\alpha) \quad . \quad (11)$$

4. DISCUSSION

Action of burning catalysts has been extensively studied (18, 19). However, the exact sites and mechanisms of catalytic action are not clear even today.   Pittman examined the effect of iron-containing catalysts in composite propellants using chemically coated AP to locate the site of catalytic action(20).   He concluded that a catalyst does not act beneath the burning surface to enhance AP decomposition or to catalyze AP-binder heterogeneous reactions.   Instead of using composite propellants, an in- teresting technique to locate the site of catalysis is the use of AP/ binder sandwiches.   The augmented burning rate with increasing pressure seems to result from acceleration of the AP decomposition reaction in the gas phase(21).   However, it would seem unrealistic to discount the con- densed phase decomposition and the role of catalysts there.   Direct evi- dence of heat release in the condensed phase, as compared with heat re- lease in the gas phase, is difficult to establish.   But Bobolev et al. showed a substantial condensed phase heat release for AP(22).   Numerous differential thermal analysis studies support the existence of enhanced decomposition rate in the condensed phase.   The slow decomposition under the conditions of thermal analysis can not be applied to deflag- ration without further consideration. However, thermal analysis is power- ful technique for screening effective catalysts in propellant combustion. Thus, it is assumed in this study that the decomposition rates in the condensed phase and the gas phase are accelerated by the addition of burning rate catalysts.

The activity of catalysts must be proportional to their concentration (X). The augmentation in the condensed phase reaction rate is written by

$$\hat{\omega}_c = 1 + C_c X ,\qquad (12)$$

where C_c is a constant representing the activity of catalysts.   It is reasonable to assume that the catalytic effect in the gas phase is gov- erned kinetically in lower pressure and diffusion-controlled in higher pressure.   Therefore, the increase in the gas phase reaction rate by catalysts is dependent on pressure (P) and given by

$$\hat{\omega}_g = 1 + C_g P^n X ,\qquad (13)$$

where C_g is another proportional constant.   In the kinetic regime, the pressure exponent must be unity.   The residence time of catalyst particles in the gaseous flame is the controlling factor in the diffusion regime. Because the residence time in the gaseous flame is proportional to P^{-2n_0}, the pressure exponent for activity of catalysts must be $1-2n_0$ where n_0 denotes the pressure exponent of the regression rate(typically 0.7 for AP). The diffusion coefficient of the gas has been assumed to be proportional to the specific volume of the gas. Thus, the pressure exponent(n) varies with pressure from 1 to -0.4. Using these results, the relative sensitivity of reaction rates for catalysts is given by

$$\hat{\omega}_c/\hat{\omega}_g = (1 + C_c X)/(1 + C_g P^n X) \qquad (14)$$

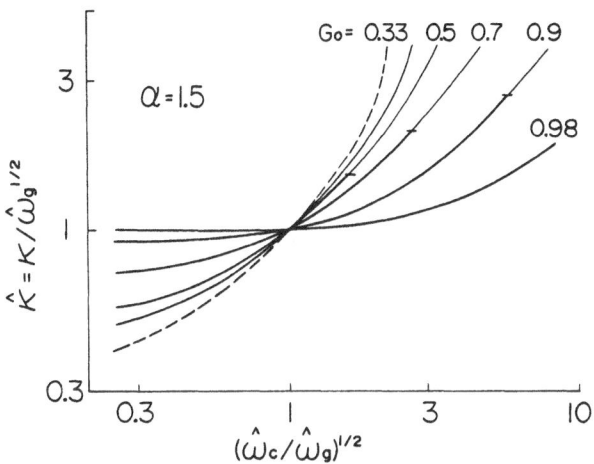

Figure 2: Relations between the augmented regression rate ($\hat{K}$) and the relative sensitivity between the reaction rates in the condensed phase and the gas phase ($\hat{\omega}_c / \hat{\omega}_g$).

Figure 2 illustrates relations between the parameter($\hat{K}$) and the relative sensitivity($\hat{\omega}_c/\hat{\omega}_g$). After specifying G_0 for an uncatalyzed AP, solutions move on the G_0 curve with the addition of catalysts. When the deflagration is solely controlled by the gas phase reaction, the burning rate of AP does not depend on the augmented rate of the condensed phase reaction (a limiting case of the gas phase deflagration model, see G_0 = 0.98 in Fig. 2). Figure 2 shows that the burning rate of the gas phase model proportionally increases with $\hat{\omega}_g^{1/2}$ as expected. On the other hand, the gradients of G_0 curves become steeper with an increase in the fraction of AP consumed in the condensed phase(decreasing G_0). This implies that the contribution of the condensed phase reaction becomes significant as a rate determining process in AP deflagration. The modi-

fication of burning rate of monopropellants caused by any variations of
reaction rates in the condensed and the gas phases can be discussed
using Fig. 2.

The regression rate of AP can be augmented by an increased reaction
rate in the condensed phase.   However, the increase of regression rate
has a stability limit described in the previous section.   The adiabatic
limits for the G_0 curves are indicated by bars in Fig. 2, above which
the deflagration wave in the condensed phase becomes unstable.   Equation
14 shows that the condensed phase reactions are enhanced by catalysts at
lower pressures. The augmented condensed phase reaction destabilizes the
deflagration because solution moves upward along G_0 curves in Fig. 2.
Increasing pressure with a fixed concentration of catalysts stabilizes
the deflagration wave since the parameter ($\hat{\omega}_c / \hat{\omega}_g$) decreases proportionally with P^{-1} .   Thus, catalyzed AP can burn self-sustainingly in
higher pressure since the gas phase reactions can follow the condensed
phase reaction enhanced by catalysts.

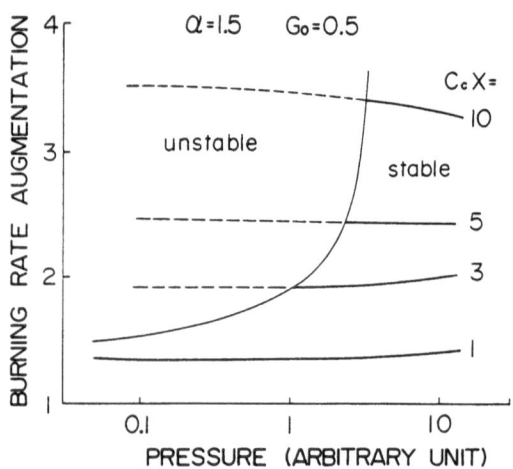

Figure 3:   Variations of regression rate augmentations(κ)
with pressure.   The activity of catalysts in the condensed
phase is denoted by $C_c X$.

The catalyzed AP can burn with an enhanced rate above the LDP.   A
typical result is illustrated in Fig. 3, where the calculation was done
assuming $\alpha = 1.5$, $G_0 = 0.5$ and $C_g / C_c = 0.1$ in the kinetic regime for
gaseous catalytic action.   Figure 3 shows that the unstable region is

extended and that the LDP increases with the catalytic activity ($C_c X$). The regression rates augmented by catalysts can be observed only in the stable region above the LDP. This result successfully reproduces the burning behavior observed in K^+ doped AP(Fig. 4 in Ref. 8). Recently, Krishnan and Periasamy studied low pressure burning of catalyzed AP-composite propellants(23). They found that the LDP of the composite propellants increases with the catalyst concentration(see Fig. 1 in Ref.23).

Effects of preheating (6,24), external radiant heat (6), external convective heat (10, 11) and the addition of fuel (12) can be included in this model. The surface temperature of burning AP weakly depends on these additions of energy because of the equilibrium vaporization at the surface. The additions of heat raise the reaction temperature predomi-nantly at the gaseous flame and promote the gaseous reaction(increasing ω_g). Catalysts added to composite propellants can modify the burning rates under the stabilizing effect of diffusion flame even below the LDP of catalyzed AP.

5. CONCLUSIONS

This study has revealed that an inherent instability of deflagration waves is responsible for the irregular behavior of the LDP of AP with catalysts. Such instability is particular to the condensed phase def-lagrations. It occurs (a) when the gaseous reaction can not catch up with the condensed phase reactions under low pressure because of its strong pressure dependence, or (b) when the condensed phase reaction is overdriven by catalysts and the gaseous flame cannot follow the enhanced condensed phase reaction. The adiabatic condition behind the condensed phase reaction zone is identified as the LDP of AP in the model. The model successfully explains the variations of the LDP and regression rate of AP with the addition of catalysts.

The surface temperature (the reaction temperature of the condensed reaction zone) is governed by an equilibrium vaporization process so that the surface temperature is not strongly dependent on the addition of ex-ternal heat. Lowered LDP due to external heat flux, preheating or the addition of fuel is found to result from enhanced gaseous flame because of the increased reaction temperature in the gas phase.

NOMENCLATURE

C_c constant representing the catalysis in the condensed phase
C_g constant representing the catalysis in the gas phase
c specific heat of condensed phase AP
G mass flux fraction of AP consumed in the gas phase flame
l_v latent heat of AP
$\dot{m}$ mass burning rate of AP
n pressure exponent of catalysis in the gas phase
n_0 pressure exponent of burning rate of neat AP
P pressure
q_c heat released in the condensed phase reaction
q_g heat released in the gas phase reaction
q_s effective heat released at the burning surface of AP(see Eq. 2)
r regression rate of AP
T temperature
T_s surface temperature of AP
T_0 initial temperature of AP
X catalyst content in AP
x streamwise coordinate

GREEK SYMBOLS

α nondimensional heat of the condensed phase reaction(see Eq. 4)
κ burning rate augmentation($=r/r_0$)
$\hat{\kappa}$ parameter denoting the contribution of the gas phase reaction to
           the modification on regression rate of AP(see Eq. 9)
λ_g thermal conductivity of gas
ω reaction rate
$\hat{\omega}_c$ reaction rate augmentation in the condensed phase by catalysts
$\hat{\omega}_g$ reaction rate augmentation in the gas phase by catalysts

SUBSCRIPTS

c condensed phase
g gas phase
0 catalyst-free

REFERENCES

1. Spalding, D.B., Combustion and Flame, vol.4, 1960, pp.59-76.
2. Steinz, J. A. and Summerfield, M., Princeton Univ., Dept. of Aerospace and Mechanical Eng., Report No.446q, 1965.
3. Johnson, W. F. and Nachbar, W., Eighth Symp.(Int.) on Combustion, Williams and Wilkins, Baltimore, 1962, pp. 678-689.
4. Hightower, J.D. and Price, E.W., Astronautica Acta, vol. 14, 1968, pp.11-21.
5. Guirao, C. and Williams, F.A., AIAA J., vol. 9, No.7, 1971, pp. 1345-1356.
6. Friedman, R., Nagent, R.G., Rumbel, K.E. and Sairlock, A.C., Sixth Symp.(Int.) on Combustion, Reinhold, New York, 1957, pp. 612-618. Williams and Wilkins, Boltimore, 1962, pp. 663-672.
7. Shadman-Yazdi, F. and Petersen, E.E., Combustion Science and Technology, vol. 5, 1972, pp. 61-67.
8. Boggs, T.L., Price, E.W. and Zurn, D.E., Thirteenth Symp. (Int.) on Combustion, Comb. Inst., Pittsburgh, 1971, pp.995-1008.
9. McHale, E.T. and von Elbe, G., Combustion Science and Technology, vol.2, 1970, pp.227-237.
10. Mitani, T. and Niioka, T., Twentieth Symp.(Int.) on Combustion, Comb. Inst., Pittsburgh, 1984, pp.2043-2049.
11. Wiersma, S.J. and Wise, H., Combustion Science and Thechnology, 19, 1978, pp.1-11.
12. Arden, E.A., Powling, J. and Smith, W.A.W., Combustion and Flame, vol. 6, 1962, pp.21-33.
13. Ben-Reuven, M. and Summerfield, M., ARBRL-CR-00507, U.S.Army Ballistic Research Lab. 1983.
14. Denison, M.R. and Baum, E., American Rocket Society Journal, vol. 31, 1961, pp.1112-1122. also see Williams, F.A., Combustion Theory, 2nd Ed., Benjamin/Cummings Publishing Co., Calif., 1985, p328.
15. Shkadinskii, K.G., Khaikin, O.I. and Librovich, V.B., Moscow, Nauka, 1975.
16. Merzhanov, A.G. and Porovinskaya, I.P., Combustion Science and Technology, vol. 10, 1975, pp.195-201.
17. Mitani, T. and Williams, F.A., Twenty first Symp.(Int.) on Combustion, Comb. Inst., Pittsburgh, 1986, pp.   . also see Sandia National Lab. Report SAND86-8230, Livermore, Calif., 1986.
18. Hall, A.R. and Pearson, G.S., Oxidation and Combustion Review, (Tipper, C.F.H. Ed.), Elsevier Publishing Co., Amsterdam, 3, No. 2, 1968, pp.129-239.
19. Kishor, K. and Sunitha, M.R., AIAA J., vol.17, No.10, 1979, pp.1118-1125.
20. Pittman Jr. C.U., AIAA J., vol.7, No. 2 1969, pp.328-334.
21. Handley, J.C. and Strahle, W., AIAA J., vol.13, No. 1, 1975, pp.5-6.
22. Bobolev, V.K., Glaskova, A.P., Zenin, A.A. and Leipunskii, O.I., Doklady Akad. Nauk, SSSR., vol. 151, 1963, p503.
23. Krishnan, S. and Periasamy, C., AIAA J., vol.24, No.10, 1986, pp. 1670-1675.
24. Cohen Nir, E., La Rechorche Aerosptiale, vol. 4, No. 2, 1972, pp.75-84.

3.   Flame Modeling Relevant to Turbulent Combustion

AN EXPERIMENTAL STUDY OF TUBULAR FLAMES IN ROTATING
AND NON-ROTATING STRETCHED FLOW FIELDS

Satoru Ishizuka
Department of Mechanical Engineering
Saitama Institute of Technology
1690 Fusaiji, Okabe, Saitama, Japan

1. Introduction

Since any fow can be locally resolved into a uniform translation, a rigid-body
rotation and a straining motion [1], it is worthy for our understanding of turbulent
combustion from a microscopic standpoint to study the behavior of flames in typical
rotating and/or straining flow fields.

For instance, it is considered that a turbulent combustion may occur within
randomly distributed vortex tubes of the diameter of Kolmogorov microscale, which is
subjected to stretching by vortices of the Taylor microscale [2]. Also, a model is
proposed for the combustion in spark-ignition engines, in which is assumed an instantaneous burning on the highly dissipative region of Kolmogorov microscale, followed
by a laminar fashion combustion proceeded outside the vortices [3]. Such a rapid
flame propagation has been really observed in a vortex ring of propane/air mixtures
and the flame velocity is found to attain a speed of about 14 m/sec [4]. A recent
experiment has also shown that a flame can propagate into a rotating combustible
mixtures at a speed about ten times as high as the burning velocity [5]. Since such
a fast combustion is followed by a reaction around the vortices, it may be important
to study the flame behavior in such a tubular geometry. Furthermore, it may be
important to emphasize that the small scale structure of turbulence is small in only
one or two directions, not in the third, hence, typically ribbons or tubes [2, 6].

Here, we shall present some experimental results concerning the behavior of
premixed flames in rotating and non-rotating, axi-symmetric stretched flow fields,
both of which have not attracted much attention until recently.

2. Experimental

Two types of burners are used in this study. One is a swirl type burner and the
other is a counterflow type burner as shown in Figs. 1a and 1b, respectively [7]. The
swirl type burner consists of a cylindrical glass tube of 13.4 mm inner diameter and
120 mm long, with a tangential inlet slit of 3 mm width. Combustible mixtures are
introduced tangentially from the inlet slit and the burned gas exits from the both
ends of the burner. The counterflow type burner consists of a porous cylinder of 30
mm inner diameter and 120 mm long. The combustible mixtures are normally ejected
from the inner surface of the burner and the burned gas exits from both ends of the
burner. Therefore, the former offers a rotating, axi-symmetric stretched flow field,

whereas the latter a non-rotating, axi-symmetric stretched flow field. The fuels used are methane, propane and hydrogen. For rich mixtures, nitrogen is ejected from both sides of the burners. The details of the burners can be found in [7] - [11].

3. Results

Most of the results have been already published in [7] - [11]. Some of these and additional results are shown in Figs. 2 - 14.

3 - 1. Swirl Type Burner

There are two parameters which define the phenomena. One is the fuel concentration Ω (% fuel) of the mixture and the other is the mean tangential ejection velocity V_t (m/sec), which is obtained by dividing the mixture flow rate by the exit area of the tangential slit.

When the ejection velocity is small, a curved flame is formed inside the burner (Fig. 2a). This flame is essentially stabilized at the rim of the tangential slit, and the luminous flame front is just developed downstream along the cylindrical wall; hence, the flame is curved.

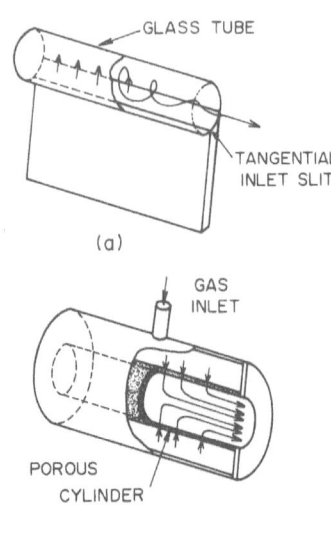

(a)

(b)

Fig. 1 Schematic of (a) the swirl type burner and (b) the counterflow type burner.

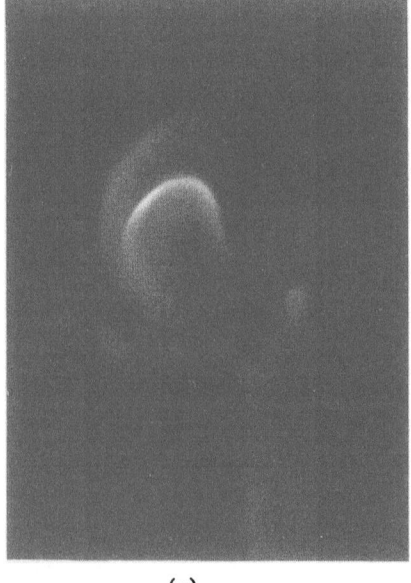

(a)

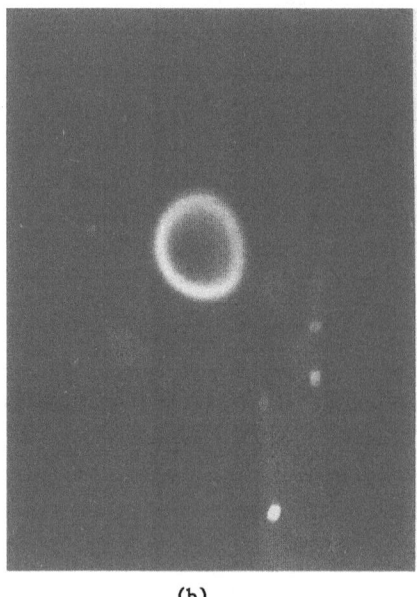

(b)

Fig. 2 Side views of (a) the curved flame ($\Omega = 6.8\%$, $V_t = 1.2$ m/sec) and (b) the tubular flame ($\Omega = 5.8\%$, $V_t = 3.0$ m/sec) of lean methane/air mixtures.

On the other hand, when the ejection velocity is large, a tubular flame with a circular cross section is established inside the burner (Fig. 2a). This flame is stabilized in a rotating flow field, and hence, the stabilizing mechanism is essentially different from that of the curved flame. Figure 3a shows the front and side views of the tubular flame of a lean methane/air mixture. The flame front is very uniform and the diameter is almost constant except at both ends, where the flame is tulip shaped due to the swirling effect. With increasing the fuel concentration, the diameter of the flame is increased and an unsteady corrugation appears at the flame front. Figure 3b show the corrugated tubular flame of a rich methane/air mixture.

Figure 4 shows the mapping of the various flame configurations for lean methane/ air mixtures. Below a critical ejection velocity V_c of 1.5 m/sec, a tubular flame is not formed. This critical velocity can be roughly estimated, because the stabilizing mechanism of the curved flame is essentially the same as the conical flame stabilized on a tube mouth. That is, when the conical flame tip is extended along the glass tube wall, makes a turn and reaches the flame base, and furthermore, the envelope curve between the base and the tip becomes tubular, then it is reasonable to consider that a tubular flame is established. This critical velocity V_c is given as 18·Su (Su : Burning velocity of the mixture) in the present burner geometry. As the burning velocity of the mixture is about 7 cm/sec according to the recent experiment [12], this estimated value (1.26 m/sec) agrees well with the observed critical velocity within 16 %. When the ejection velocity exceeds the critical velocity, a tubular flame is formed. The fuel concentration at extinction jumps down to a smaller value once the tubular flame is formed, and the values are almost constant of 5.3 % over velocities, which is close to the lean flammability limit. On the other hand, the fuel concentration at the boundary of the corrugated tubular flame is increased with increasing the ejection velocity.

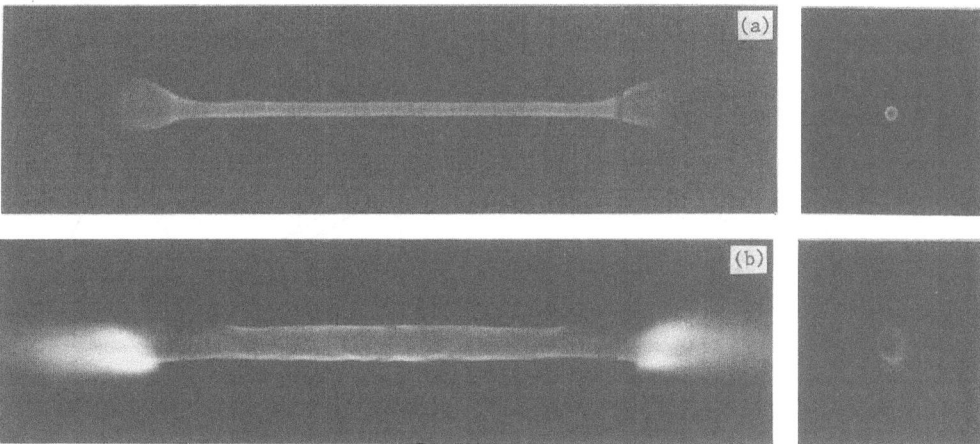

Fig. 3 Appearance of the tubular flames of methane/air mixtures with
(a) a uniform flame front ($\Omega = 6.0\,\%$, $V_t = 5$ m/sec) and with (b)
a corrugated flame front ($\Omega = 12.6\,\%$, $V_t = 5$ m/sce).

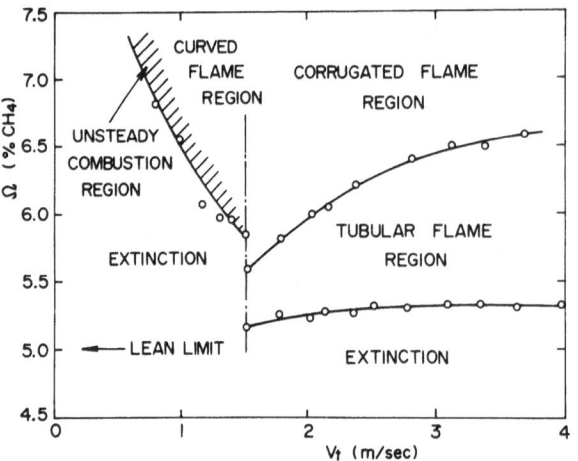

Fig. 4 Mapping of the various flame configurations of lean methane/air mixtures.

Figure 5 shows the temperature distributions across the tubular flames in the radial direction. The combustion field is separated into an outer region of low temperature and an inner region of high temperature. Therefore, it is considered that an unburned gas comes from outside with rotation, passes through the flame zone, and the burned gas is exhausted from the both ends of the axis of rotation. This structure of an inner hot gas core of low density and a surrounding unburned mixture of high density is aerodynamically stable according to the Rayleigh stability criterion [13]. Furthermore, as the temperature distribution behind the flame is symmetrical, there seems almost no heat loss behind the flame except a radiative loss [14]. Hence, the tubular flame is, also, thermally stable.

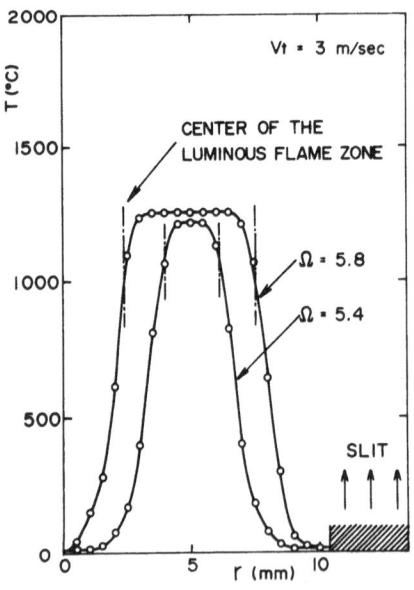

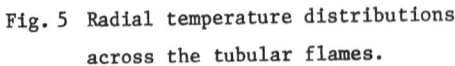

Fig. 5 Radial temperature distributions across the tubular flames.

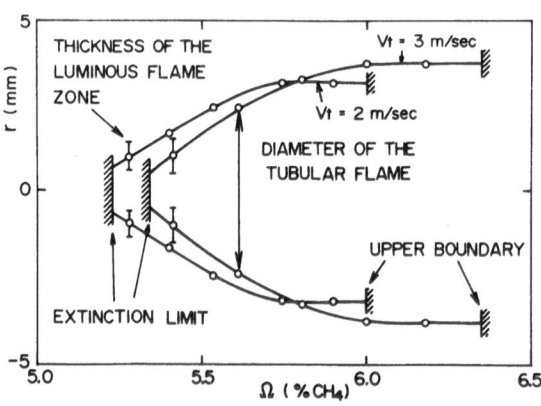

Fig. 6 Variations of the flame diameter with the fuel concentration.

Figure 6 shows the variations of the flame diameter with the fuel concentration at constant ejection velocities of 2 m/sec and 3 m/sec. With a decrease of the fuel concentration, the flame diameter is decreased and the flame is extinguished when the diameter is almost the same order as the thickness of the luminous flame zone. On the other hand, with an increase of the fuel concentration, the flame diameter is increased and the flame front becomes corrugated.

Further experiments have been performed for various mixtures. An important point is that the tubular flame is essentially established in a stretched flow field, and hence, the characteristics of the flame are very sensitive to the Lewis number of the limiting reactant as the other stretched flames [15 - 19]. Figures 7 - 9 show the mapping of the various flame configurations and the variations of the flame diameter with the fuel concentration for methane, propane and lean hydrogen mixtures.

For the mixtures near the stoichiometry, a tubular flame is not formed for the present velocities up to 6 m/sec, while, for the mixtures apart from the stoichiometry, a tubular flame is formed and its characteristics are mainly controlled by the Lewis number of a limiting (deficient) reactant. For the mixture of $Le \leq 1$ ($Le \equiv \kappa/D$, κ: thermal diffusivity, D: mass diffusivity), say, lean methane, rich propane or lean hydrogen mixtures, a tubular flame with a uniform flame front is formed. With varying the fuel concentration away from the stoichiometry, the diameter is decreased and the flame is eventually extinguished when the flame is almost attached (lean methane) or merged into a luminous rod (rich propane and lean hydrogen). The fuel concentrations at extinction are close to the flammability values. On the other hand, for the mixtures of $Le > 1$, say, rich methane or lean propane mixtures, the flame diameter cannot become small and the flame front is always corrugated. The fuel concentrations at extinction are not close to the flammability values, which is because the flame suffers from various degrees of stretch in various directions once the flame loses its cylindrical symmetry [7, 11, 20].

3 - 2. Counterflow Type Burner

Since there is no centrifugal force of rotation, the flame is strongly influenced by the gravitational force.

When the burner is mounted horizontally, the flame loses its cylindrical symmetry. For lean methane/air mixtures, the bottom part of the flame is pushed upward and the flame becomes mashroom shaped as shown in Fig. 10. However, for lean propane /air mixtures, the bottom part of the flame is stiff for bending and intensified for burning as shown in Fig. 11. For lean hydrogen/air mixtures, the bottom part of the flame is extinguished (Fig. 12a), which is probably due to the preferential diffusion of hydrogen species as in the case of tip opening or cellular flame. With further decreasing the fuel concentration, a pair of tubular flames is formed as shown in Fig. 12b. The differences in these flame behaviors are attributed mainly to the different Lewis numbers of the mixtures.

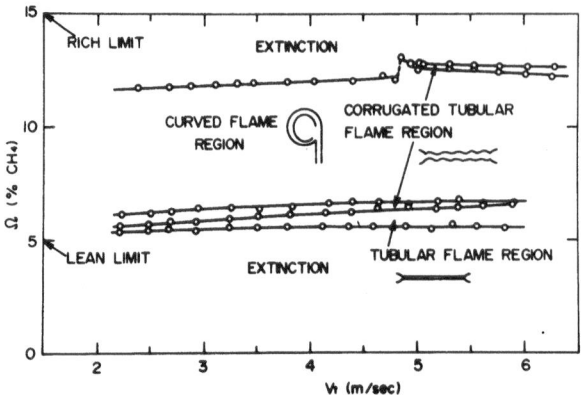

Fig. 7(a)   Mapping of the various flame configurations of methane/air mixtures.

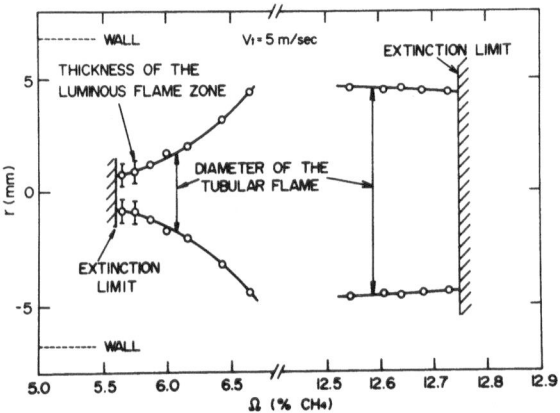

Fig. 7(b)   Variations of the diameter of the tubular flame of methane/air mixtures.

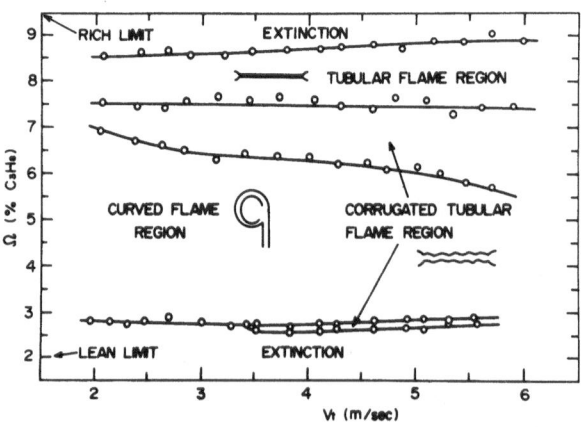

Fig. 8(a)   Mapping of the various flame configurations of propane/air mixtures.

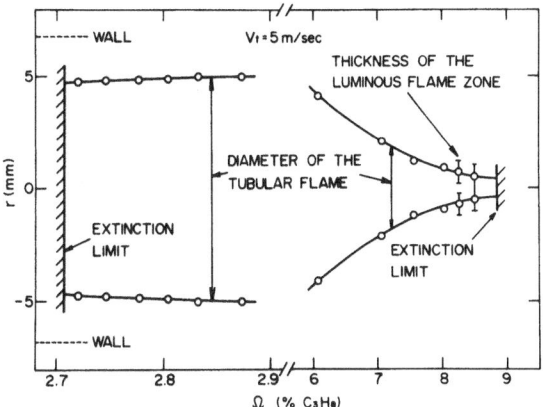

Fig. 8(b) Variations of the diameter of the tubular flame of propane/air mixtures.

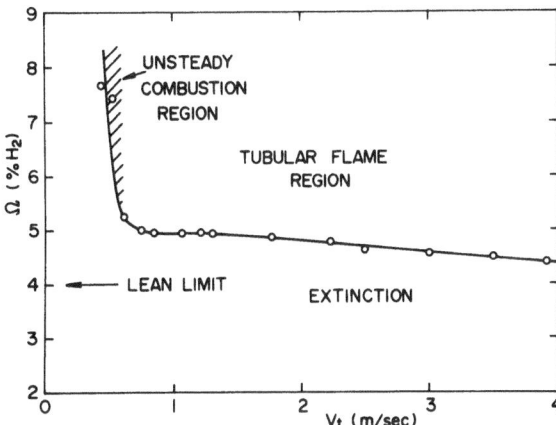

Fig. 9(a) Mapping of the various flame configurations of lean hydrogen/air mixtures.

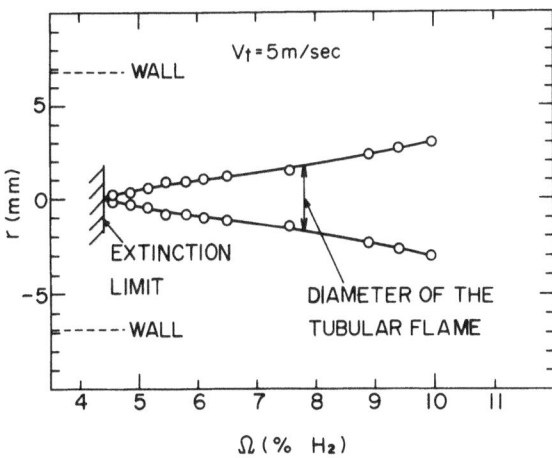

Fig. 9(b) Variations of the diameter of the tubular flame of lean hydrogen/air mixtures.

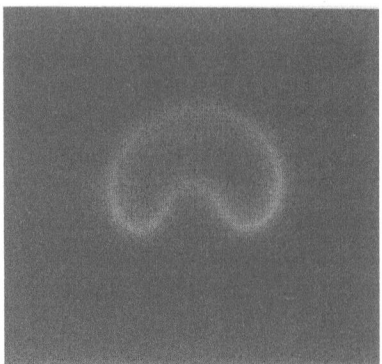

Fig. 10   Appearance of a lean methane/
air flame in the horizontally mounted
counterflow type burner ($\Omega = 5.3$ %, $V_n =$
10 cm/sec).

Fig. 11   Appearance of a lean propane/
air flame in the horizontally mounted
counterflow type burner ($\Omega = 2.1$ %, $V_n =$
5.8 cm/sec).

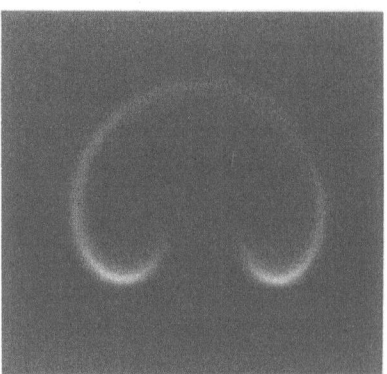

Fig. 12(a)   Appearance of a lean
hydrogen/air flame in the horizontally
mounted counterflow type burner ($\Omega = 6.1$
%, $V_n = 2.4$ cm/sec, CH_4; added).

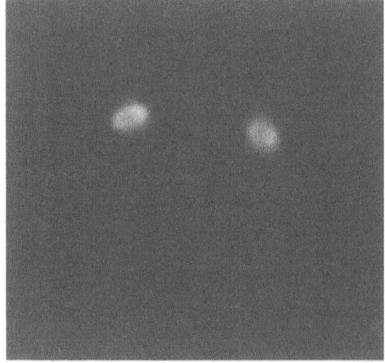

Fig. 12(b)   Appearance of a lean
hydrogen/air flame in the horizontally
mounted counterflow type burner ($\Omega = 3.8$
%, $V_n = 2.3$ cm/sec, CH_4; added).

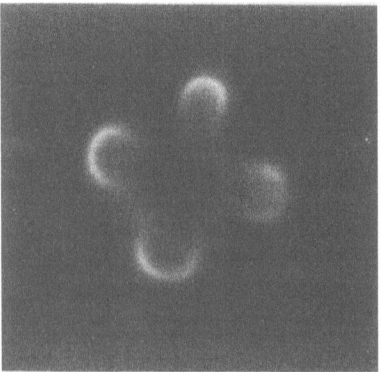

Fig. 13(a)   Appearance of a lean
hydrogen/air flame in the vertically
mounted counterflow type burner ($\Omega = 7.1$
%, $V_n = 5.7$ cm/sec, CH_4; added).

Fig. 13(b)   Appearance of a lean
hydrogen/air flame in the vertically
mounted counterflow type burner ($\Omega = 6.0$
%, $V_n = 11$ cm/sec).

On the other hand, when the burner is mounted vertically, the flame usually holds its cylindrical symmetry, although the flame diameter slightly varies from the bottom to the top. However, for lean hydrogen or rich propane/air mixtures, a non-uniform structure appears at the flame front when the flame diameter is large. Figure 13a shows the side view of the flame of lean hydrogen mixtures. The shape is polypetalous. The number of the petal is decreased with decreasing the fuel concentration, and eventually the flame becomes tubular near the extinction limit. The variations of the flame diameter with the fuel concentration for three lean mixtures are shown in Fig. 14, in which the diameter of lean hydrogen mixtures around 6 % is averaged because the flame shape is not cylindrical as shown in Fig. 13b. For lean methane or hydrogen mixtures the diameter

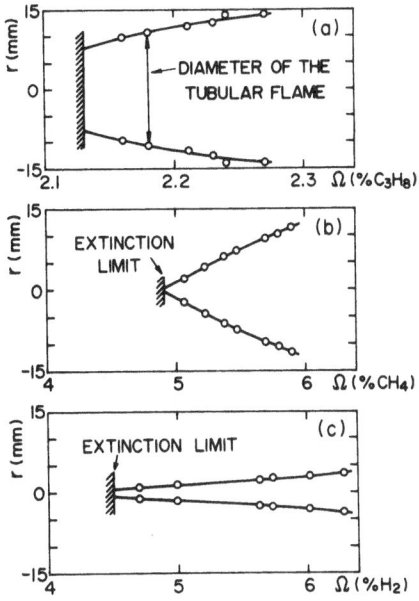

Fig. 14 Variations of the tubular flame diameter in the vertically mounted counterflow type burner.

can become small, but for lean propane mixtures, not. These flame behaviors agree qualitatively with those in the other stretched flow fields [15 - 19].

4. Discussion and Concluding Remarks

The distinct behaviors of flames in the rotating flow field have been understood on the basis of those in the non-rotating flow field obtained by the vertically mounted counterflow type burner and the tangential velocity distribution of the Rankine's combined vortex [7]. Namely, for the mixtures of Le ≤ 1, a flame can be established near the stagnation plane (here, the axis of rotation). Hence, the diameter can be small and the flame exists within a forced vortex region of a solid-body rotation. The flame front is not deformed and the flame can survive near the flammability limit. On the other hand, for the mixtures of Le > 1, the flame cannot be established near the stagnation plane. Hence, the diameter cannot become small and the flame is situated in the free vortex region of angular deformation. However, this may not directly result in the corrugation of the flame front, because a recent analysis shows that the tangential velocity distribution does not influence the flame characteristics as long as a cylindrical symmetry is assumed [21]. Furthermore, it is shown that the cellular instability is suppressed in the presence of rotation [22]. For the time being, the possible origines for the corrugation seem to be (a) non-symmetrical deformation due to the existence of the tangential slit, (b) the radial velocity distribution with a peak value [7] and (c) the Taylor - Görtler instability along the glass wall [11]. Of course, the flame located near the axis of rotation may be more stable because the forced vortex flow near the axis is stable whereas the free vortex flow is only neutral according to the Rayleigh stability criterion, and because the flame near the stagnation axis suffers from a more stretch which has a stabilizing effect for the cellular instability [19].

After all, the tubular flame with a uniform flame front can be established for the mixtures of Le $\leq$ 1 in the rotating and non-rotating flow fields as long as the diameter is small, and for the mixtures of Le > 1 in the non-rotating flow field, provided that the body force evenly works on the flame surface.

An interesting case is the formation of a pair of tubular flames in lean hydrogen mixtures (Fig. 12). In this case, the flame - flow interaction together with a buoyant force probably makes a pair of vortices, resulting a formation of a pair of tubular flames in the rotating flow field. The establishment of this mechanism may depend on the Lewis number of the mixture. Hence, it is suggested that the possibility of the formation of tubular flames in a turbulent flow will be higher as the Lewis number of the mixture becomes smaller.

[References]
1. Batchelor, G. K.: An Introduction to Fluid Mechanics, p.79/84, Cambridge University Press, 1967.
2. Chomiak, J.: Sixteenth Symposium (International) on Combustion, p.1665, The Combustion Institute, 1977.
3. Tabaczynski, R. J., Triker, F. H., and Shannon, B. A.: Combustion and Flame 39, 111(1980).
4. McCormack, P. D., Scheller, K., Mueller, G., and Tisher, R.: Combustion and Flame 19, 297(1972).
5. Ishizuka, S.: Twenty-fifth Japanese Symposium on Combustion, p.169, 1987.
6. Kuo, A. Y. and Corrsin, S.: Journal of Fluid Mechanics 50, 285(1971).
7. Ishizuka, S.: Twentieth Symposium (International) on Combustion, p.287, The Combustion Institute, 1984.
8. Ishizuka, S.: Twenty-second Japanese Symposium on Combustion, p.127, 1984.
9. Ishizuka, S.: Twenty-third Japanese Symposium on Combustion, p.397, 1985.
10. Ishizuka, S.: Twenty-fourth Japanese Symposium on Combustion, p.76, 1986.
11. Ishizuka, S.: An experimental study on extinction and stability of tubular flames, presented at the joint technical meeting of the western states section and the Japanese section of the Combustion Institute, Hawaii, November, 1987.
12. Yamaoka, I. and Tsuji, H.: Twentieth Symposium (International) on Combustion, p.1883, The Combustion Institute, 1984.
13. Lord Rayleigh: Proc. Roy. Soc. (London) A93, 148(1916).
14. Liu, G. E., Ye, Z. Y., and Sohrab, S. H.: Combustion and Flame 64, 193(1986).
15. Tsuji, H. and Yamaoka, I.: Nineteenth Symposium (International) on Combustion, p.1583, The Combustion Institute, 1982.
16. Sato, J.: Nineteenth Symposium (International) on Combustion, p.1541, The Combustion Institute.
17. Ishizuka, S and Law, C. K.: Nineteenth Symposium (International) on Combustion, p. 327, The Combustion Institute, 1982.
18. Law, C. K., Ishizuka, S., and Mizomoto, M.: Eighteenth Symposium (International) on Combustion, p.1791, The Combustion Institute, 1981.
19. Ishizuka, S., Miyasaka, K., and Law, C. K.: Combustion and Flame 45, 293(1982).
20. Matalon, M.: Combustion Science and Technology 29, 225(1983).
21. Takeno, T. and Ishizuka, S.: Combustion and Flame 64, 83(1986).
22. Sivashinsky, G. I., Rakib, Z., Matalon, M., and Sohrab, S. H.: Flame propagation in a rotating gas, to appear.

BURNING VELOCITY OF STRETCHED FLAMES

Tadao TAKENO, Makihito NISHIOKA and Satoru ISHIZUKA*
Institute of Interdisciplinary Research, The University of Tokyo,
Komaba, Meguro-ku, Tokyo 153, Japan.
* Saitama Institute of Technology, 1690 Fusaiji, Okabe, Saitama, Japan.

1. INTRODUCTION

The laminar flamelet concept is coming to attract attention in turbulent combustion modeling [1,2]. It points to the importance of
properties of laminar flames in various flow fields in determining
turbulent combustion rates. The asymptotic analysis is very useful
to derive the required flame properties, especially those of the
stretched flames in nonuniform flows. The analysis is often based on
the flame surface model of zero reaction zone width, as well as on
the one-step kinetics [3]. In most problems of the flame and flow
interaction, the reaction zone width is much smaller than the scale
of flow field, and hence the flame surface model has successfully
been applied. This makes the mathematical analysis extremely simple,
and the asymptotic analysis makes the most of this to derive analytical solution for burning velocity of stretched flames, which is one
of the most important flame properties. However, most of the analyses are obliged to use the simple flow solution of constant density.
Then the validity of the derived results is a matter of controversy,
in view of the large flow expansion due to heat release in the flames.
One may suspect they should not give even a qualitatively correct
picture. The survey shows that there is only one study [4], on a
plane hydrogen-air flame stabilized in a stagnation flow and subjected to weak stretch, which refers to this problem. We should be more
concerned with this problem so as to recognize the utility and limi-
tation of the simplified constant density analysis. Another controversy problem of stretched flames is the physical mechanism through
which the stretch affects the burning velocity. Although there are
some discussions on the possible mechanisms for a plane stagnation
flame [5], it seems that there has been no definite explanation yet,
which is universally applicable to any stretched flames. In this
paper we will advance some arguments on these two problems, in the
light of our recent analytical and numerical studies of the plane and

tubular flames stabilized in stagnation flows [6,7]. Although the
studies are based on the simplified one-step kinetics, they have
revealed the essential characteristics of the flame and flow interaction and the arguments will elucidate some important features of the
stretched flames.

2. DEFINITION OF BURNING VELOCITY AND STRETCH

In the stretched flames the velocity and mass flux convected normal
to the flame change with distance and hence the definition of burning
velocity becomes ambiguous. One suggestion is to take the velocity
of the upstream boundary of preheat zone, while another suggestion is
to take the velocity at the downstream boundary of reaction zone. In
the plane stagnation flame, the former has been deemed to correspond
to the local minimum velocity [5,8], whereas the latter to the local
maximum velocity [9]. The other suggestion is to take the velocity
at the plane where heat release occurs [10]. In the asymptotic
analysis based on the flame surface model, this is the reaction zone
plane of zero thickness. This definition was adopted in our previous
asymptotic analysis [11]. In the numerical solution with finite

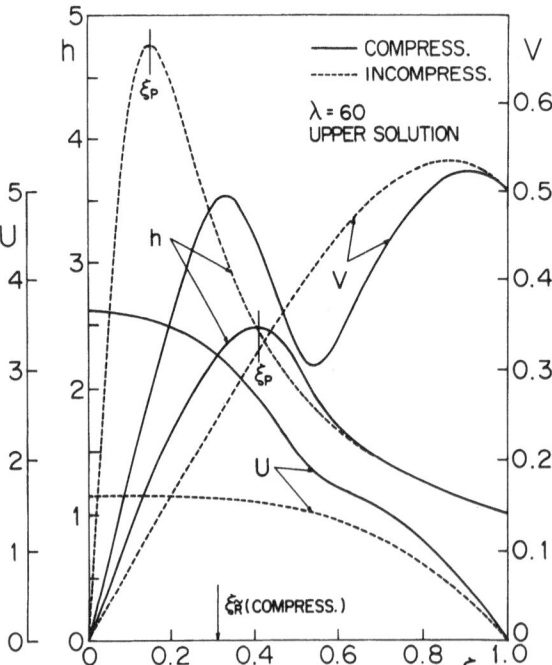

Fig. 1 Flow field of tubular
flame.

Fig. 2 Variation of normal
velocity profile with
injection velocity.

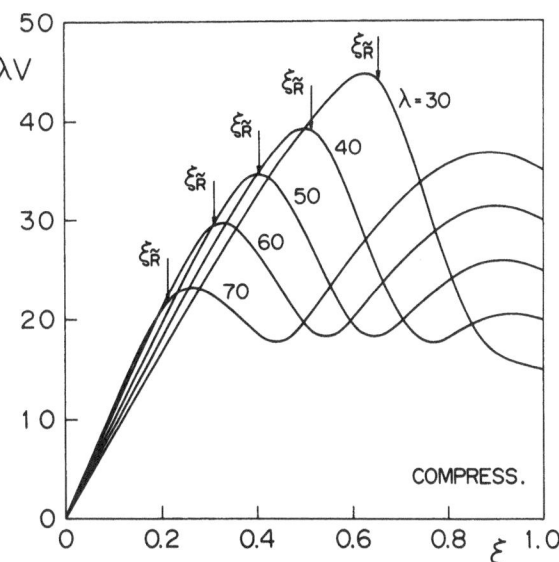

reaction zone width, on the other hand, the natural extension of this
idea is to take the plane where heat release rate becomes maximum.
This is the one adopted in our numerical study of the tubular flame,
with the incompressible flow solution, and the correlation of the
derived result with that of the asymptotic analysis was found satisfactory [11].

The definition of burning velocity becomes more complicated for
the compressible flow solution. Usually, the velocity normal to the
flame is accelerated and decelerated through the flame zone, and the
characteristic velocity profile is produced. As an example, the
calculated velocity field for the tubular flame is shown in Fig. 1
[6]. The reduced velocities normal and parallel to the flame, V and
U respectively, are plotted against the normal distance ξ, as compared to those of the incompressible flow solution. $\xi_{\tilde{R}}$ represents
the position in the flame where heat release rate becomes maximum.
The normal velocity profile changes with the injection velocity λ in
a complicated manner, as is shown in Fig. 2. However, when the
normal mass flux, instead of linear velocity, is plotted against the
distance, it increases almost linearly from the axis up to the position $\xi_{\tilde{R}}$, as is seen in Fig. 3. This suggests that we may take the
normal mass flux at $\xi_{\tilde{R}}$ as the mass burning velocity of the flame.
This definition was adopted in the study with the compressible flow
solution [6]. In addition, the burning velocity and the mass burning
velocity defined by the local minimum velocity [5,8] were calculated
as well in the present study, for the purpose of comparison.

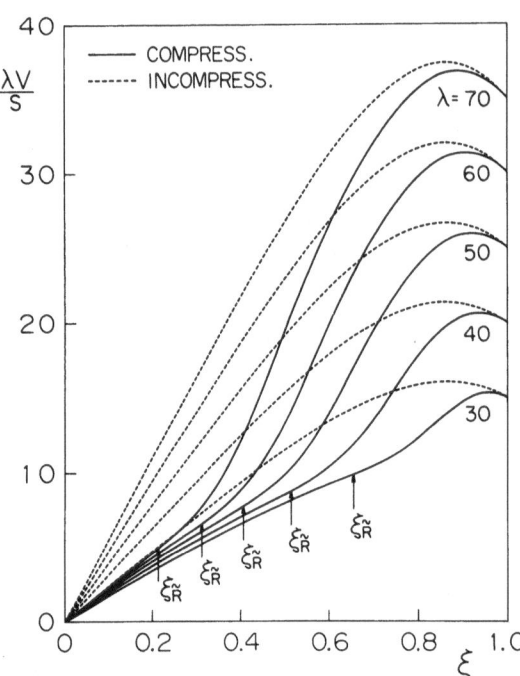

Fig. 3 Variation of normal mass flux profile with injection velocity.

The definition of stretch is another problem. In the flames stabilized in stagnation flows, the stretch is proportional to the velocity component parallel to the flame [7]. As is seen in Fig. 1, the magnitude of this component, and hence the stretch become to depend on the position where we should take the flame front. In the present paper, the positions, where we define the burning velocity or the mass burning velocity, were taken as well to define the stretch, so as to keep consistency.

3. CORRELATION OF BURNING VELOCITY WITH STRETCH

The numerical calculation was performed for lean methane-air mixture [6]. Figure 4 shows the response curves of the mass burning velocity plotted against the stretch K, defined in the way described above. The former is made nondimensional by the mass burning velocity of the normal plane flame, while the latter by the burning velocity and flame thickness of the normal plane flame [6,11]. The results predicted by the asymptotic analysis, based on the potential flow, are compared with those obtained by numerical calculations for the

Fig. 4 Burning velocities plotted against flame stretch.

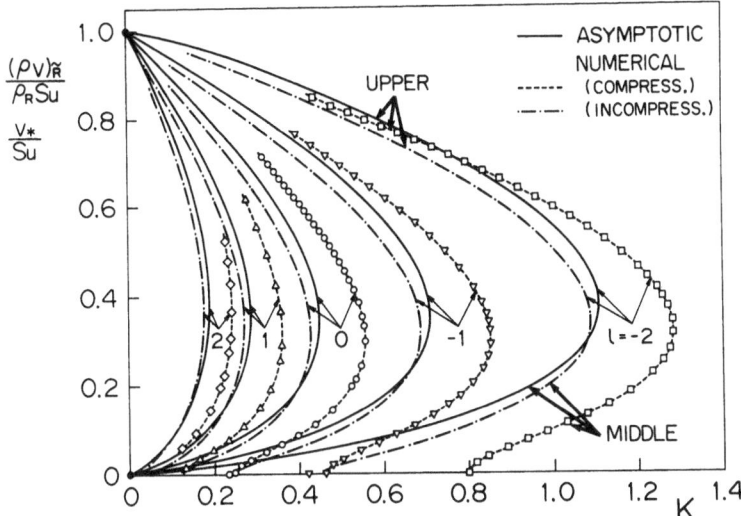

compressible viscous flows. The parameter l represents the deviation
of Lewis number from unity [11]. It can be seen that the asymptotic
analysis predicts somewhat smaller critical values of the stretch for
extinction. However, we may say that the general qualitative behavior,
especially the dependence on Lewis number, is well predicted by the
asymptotic analysis.

The above agreement is rather surprising when we consider the
substantial change induced in flow field for the compressible flow
solution. Detailed examinations of the numerical solution have revealed the following flow and flame characteristics. The flow expansion due to heat release accelerates the velocity parallel to flame,
as well as the velocity normal to flame. The former acceleration
produces a considerable increase in the flame stretch along the flame,
as compared to the incompressible flow solution of the same velocity
gradient. However, when the temperature dependence of transport
properties is taken into account, the accelerated transport processes
at the elevated temperature brings about an increase in the reaction
zone width. This eventually makes smaller the velocity gradient
across the reaction zone. That is, the stay time of gas particles in
the reaction zone is increased to weaken the stretch. This offsets
the increased stretch due to the flow expansion, and will explain the
obtained agreement.

Although this is the result for a specific flow field and the
agreement may happen by coincidence, the above explanation seems quite
general. In view of the similar agreement observed for the plane

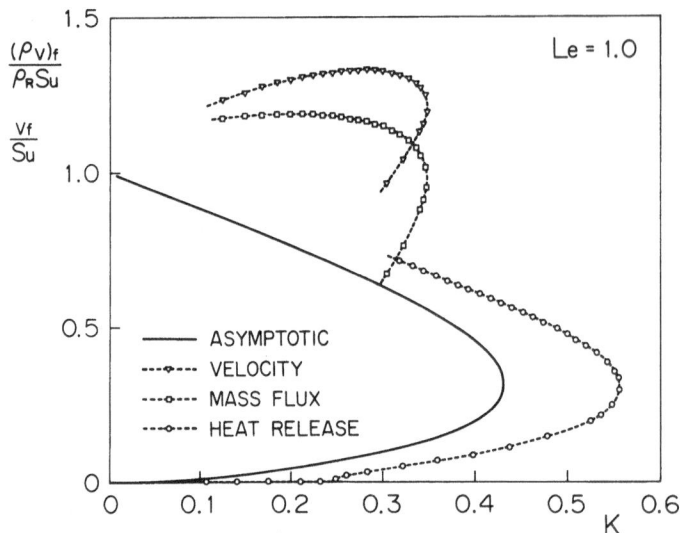

Fig. 5 Response curves for distinct definitions of flame front.

stagnation flame [4], we may hope that this fortunate offset is the universal one, which makes the asymptotic analysis more useful than had been expected.

The related problem is how the definition of burning velocity and stretch affects the response curves. Figure 5 shows the response curves calculated for the aforementioned different definitions of flame front for Lewis number $Le = 1.0$. The reduced burning velocity and mass burning velocity defined at the position of the local minimum velocity are compared with the mass burning velocity defined at $\xi_{\tilde{R}}$, as well as with the burning velocity of the asymptotic analysis. The local minimum velocity criterion predicts the larger reduced burning velocity and mass burning velocity, and the smaller stretch. Along the upper solution, the reduced burning velocity is larger than the reduced mass burning velocity, and they are always larger than unity. Moreover, the burning velocity increases with the stretch, except for the region near the extinction. This is highly improbable physically, since the stretch should decrease the burning velocity for the case of $Le = 1.0$, where the flame temperature remains almost the adiabatic flame temperature. On the other hand, the mass burning velocity defined at $\xi_{\tilde{R}}$ decreases reasonably as expected. We think that the local minimum velocity position does not represent correctly the upstream boundary of preheat zone, and hence is not adequate for definition of the burning velocity. This will explain the reason why the burning velocity, measured in the experiment by this criterion, increased with the stretch [8].

4. PHYSICAL MECHANISM THROUGH WHICH STRETCH AFFECTS BURNING VELOCITY

One decided advantage of the flame surface model is that for any flames in various flow fields we can apply the inner solution of the steady normal plane flame and we need not be involved in solving the internal flame structure. This means that we can consider the up-stream temperature gradients of these flames are just equal to that of the normal flame, irrespective of the stretch and curvature they experience. It appears that some important implications of this fact have not fully been discussed yet. In the following we can make use of this fact to explain the physical mechanism through which the stretch can affect the burning velocity.

The upper part of Fig. 6 explains the energy balance for a steady, adiabatic, plane, one-dimensional flame surface located at x_*, with the flame temperature T_* and the mass burning velocity m_*. It can be presented in the following form.

$$\{ \lambda \frac{dT}{dx} \}_{x_*} = m_* \bar{c}_p (T_* - T_u), \tag{1}$$

where T_u represents the initial temperature and we have to take some appropriately defined constant value for the specific heat at constant pressure $\bar{c}_p$. The equation just suggests that the heat flux conducted upstream from the flame surface must be equal to the thermal enthalpy flux of the product gas leaving the flame surface. If we use T_a and m to represent the adiabatic flame temperature and the mass burning velocity, respectively, of the normal plane flame, then the reduced mass burning velocity of any stretched flames with flame temperature T_* can be given as

$$\frac{(\rho v)_*}{m} = \frac{m_*}{m} \frac{(\rho v)_*}{m_*} = \frac{m_*}{m} \frac{(\rho v)_* \bar{c}_p (T_* - T_u)}{m_* \bar{c}_p (T_* - T_u)} = \frac{m_*}{m} \frac{(\rho v)_* \bar{c}_p (T_* - T_u)}{ \{ \lambda \frac{dT}{dx} \}_{x_*} }, \tag{2}$$

where $(\rho v)_*$ represents the mass flux normal to the flame surface. The last expression shows that the mass burning velocity is given by product of two terms. The first one represents the well known effect of the flame stretch, that is the change in normal burning velocity due to a variation in the flame temperature from T_a to T_*. The vari-

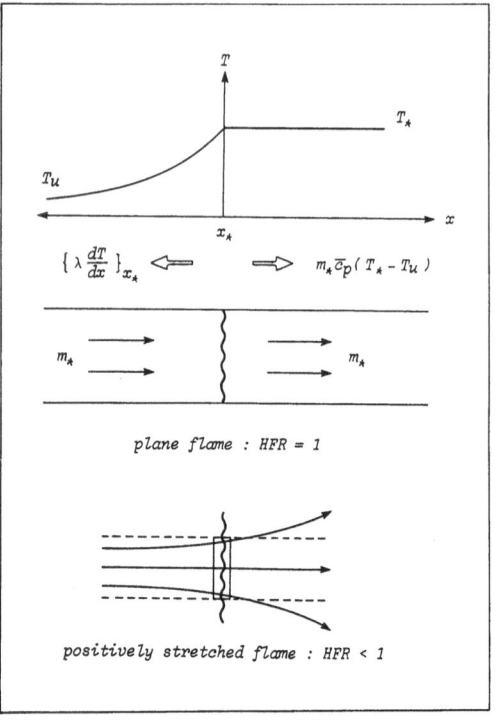

Fig. 6 Energy balance of
flame surface model.

ation is provided for any mixtures with nonunity Lewis number when-
ever the flame is subjected to the stretch, because of the unbalance
in heat and mass diffusion rate. The first term, therefore, is
rather easy to understand. On the other hand, the second term represents the fraction of the heat flux coming back to the flame. We may
call the fraction "heat feedback ratio". This is just unity for the
normal flame, of course. However, in any steady flames stabilized in
divergent flows, for example, a part of the heat flux conducted upstream is convected away by the flow component parallel to the flame
surface, and hence the heat feedback ratio inevitably becomes smaller
than unity. The lower part of Fig. 6 explains this situation. That
is, the reduced heat feedback ratio is a characteristic of the flame
stabilized in divergent flows, or of the positively stretched flame.
The above equation shows that the mass burning velocity decreases in
proportion to the reduced heat feedback ratio. On the contrary, the
heat feedback ratio will be increased over unity in the negatively
stretched flame. Then the mass burning velocity will exceed that of
the normal plane flame, for unity Lewis number.

    In this way we see that the stretch affects the mass burning
velocity through the two different physical mechanisms: the change in
flame temperature and the nonunity heat feedback ratio. This is a

direct consequence of the flame surface model, and is a universal one applicable to any steady or nonsteady adiabatic flames, with or without curvature, as long as the flame surface model remains valid.

References

1.   Peters, N.: Laminar flamelet concepts in turbulent combustion, Twenty-First Symposium (International) on Combustion, in press.

2.   Bray, K. N. C.: Methods of including realistic chemical reaction mechanisms in turbulent combustion models, Proceedings of Second Workshop on Modeling of Chemical Reaction Systems, in press.

3.   Buckmaster, J. and Ludford, G. S. S.: Theory of Laminar Flames, p. 44, Cambridge Univ. Press, 1983.

4.   Warnatz, J. and Peters, N.: Dynamics of Flames and Reactive Systems, Progress in Astronautics and Aeronautics Vol. 95, p. 61, AIAA, 1984.

5.   Wu, C. K. and Law, C. K.: Twentieth Symposium (International) on Combustion, p. 1941, The Combustion Institute, 1985.

6.   Nishioka, M.: Effects of variable density on a tubular flame (in Japanese), Ms. Thesis, Graduate School of Aeronautics, University of Tokyo (1987).

7.   Takeno, T., Ishizuka, S. and Nishioka, M.: Effects of geometrical configuration on extinction behavior of flames stabilized in stagnation flows (in preparation).

8.   Law, C. K., Zhu, D. L. and Yu, G.: Propagation and extinction of stretched premixed flames, Twenty-First Symposium (International) on Combustion, in press.

9.   Mendes-Lopes, J. M. C. and Daneshyar, H.: Combus. Flame 60, 29 (1985).

10.   Williams, F. A.: Combustion Theory, 2nd Edition, p. 418, The Benjamin/Cummings, 1985.

11.   Takeno, T. and Ishizuka, S. and Nishioka, M.: Combus. Flame 66, 271 (1986).

CONVECTION EFFECTS AND THE STABILITY OF HYDROGEN FLAME BUBBLES

J. Buckmaster, R. E. Johnson
University of Illinois,Urbana, IL 61801
and S. Weeratunga

Abstract

Hydrogen 'flame-bubbles' are the small flames observed in very lean hydrogen/air mixtures. Previous estimates of the flame temperature fail to account for convection, and a simple mathematical model is proposed to remedy this. In addition, convective gradients generated by the buoyancy of the hot gases are proposed as a mechanism to overcome the instability that appears to be associated with the large curvature of the flames.

Introduction

It is well known that very lean (but flammable) mixtures of hydrogen and air will not support a single tent-shaped flame when ignited at the bottom of a tube. Rather, a number of very small flames are generated [1], 4 mm or so in diameter, which appear to have independent existence and rise in an unchanging fashion, suggesting that each is associated with a stationary combustion field. These flames are highly curved and this plays an essential role in their existence. The flux of the highly diffusive hydrogen towards the reaction zone is enhanced by the curvature, so that flame temperatures are achieved in excess of the adiabatic flame temperature. Thus these flames are observed in mixtures that are too weak to sustain an adiabatic plane flame, mixtures, indeed, for which the adiabatic flame temperature is less than the ignition temperature.

This effect is clearly seen in a simple mathematical model due to Zeldovich, a convection-free model [2]. Consider a spherical premixed flame in which the velocity is everywhere zero. This would require a gravity-free environment, and so differs in an important way from the opening scenario, but is, nevertheless, illuminating.

Adopting the crudest of kinetic models, we have

$$0 = \frac{\lambda}{r^2} \frac{d}{dr}\left(r^2 \frac{dT}{dr}\right) + Q\,\Omega(Y,T)$$

$$0 = \frac{(\rho D)}{r^2} \frac{d}{dr}\left(r^2 \frac{dY}{dr}\right) - \Omega(Y,T)$$

(1)

where T is the temperature, Y the mass fraction of hydrogen, and Ω is the reaction rate.

Eliminating Ω and integrating,

$$T + \frac{Q}{L\,C_p}\,Y = T_f + \frac{Q}{L\,C_p}\,Y_f \tag{2}$$

where the subscript f denotes the fresh far-field state, and a term in $\frac{1}{r}$ has been dropped since it is unacceptable at the origin. If we assume that reaction ceases behind the flame because Y vanishes, it follows that the flame temperature is

$$T_* = T_f + \frac{QY_f}{C_p L} \quad , \tag{3}$$

which should be compared with the adiabatic flame temperature

$$T_a = T_f + \frac{QY_f}{C_p} \tag{4}$$

With a Lewis number (L) in the neighborhood of 0.3, T_* is significantly larger than T_a.

The same enhanced temperature can arise in hydrogen tubular flames, discussed elsewhere in these proceedings by S. Ishizuka.

If we adopt a flame-sheet model in which Ω is assumed to be a Dirac δ-function of strength $e^{-E/2RT_*}$, explicit calculation of the steady state combustion field is possible and, in addition, a stability analysis can be carried out [2]. This shows that the steady solution is unconditionally unstable, a fact that has been exploited by Joulin in his ignition studies, e.g. [3]. The instability leads either to collapse and disappearance of the flame, or growth and propagation into the surrounding mixture. This is not a Lewis number effect since instability is found for all values of L, but appears to be associated with the strong curvature. Thus Zeldovich's model, while explaining one mystery (how these flames can exist in such weak mixtures) introduces another: why are they stable?

It is conceivable (but intuitively unlikely) that a more realistic kinetic scheme would not exhibit the difficulty. Alternatively, the convective flux generated by buoyancy forces might provide a stability mechanism, and we will discuss this possibility together with other important convective effects.

The Flame-Cap Model

In the absence of gravity it is natural to look for a symmetrical solution in which the flame sheet forms a closed surface. For very weak gravitational fields, where only weak convective currents will be generated, the sheet is likely to remain closed. But for larger rise speeds it is reasonable to speculate that the cooling at the rear of such a sheet, generated by the adverse convective flux, will cause local extinction. This is consistent with the appearance of the physical flames, although that can only be suggestive. Be that as it may, we shall consider a model in which the flame consists of a cap, convex when viewed from above (Fig. 1). Extinction at the rim of the cap can arise from a decrease in the convective flux of hydrogen with increasing σ, or a decrease in the local

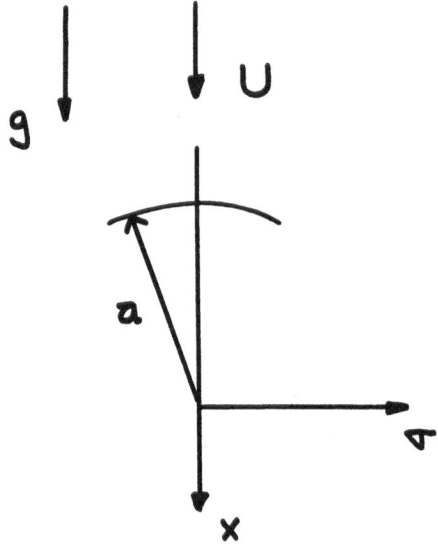

Fig. 1. The Flame-Cap Model

curvature which diminishes the enhancement identified in the earlier section. However, this plays no role in our analysis.

We will consider simple model equations that are valid only on the axis of symmetry σ = 0. These are

$$C_p \, \rho u \left. \frac{\partial T}{\partial x} \right|_{\sigma = 0} = \lambda \nabla^2 T \Big|_{\sigma = 0} + Q\Omega \Big|_{\sigma = 0} \; ,$$

$$\rho u \left. \frac{\partial Y}{\partial x} \right|_{\sigma = 0} = (\rho D) \, \nabla^2 Y \Big|_{\sigma = 0} - \Omega \Big|_{\sigma = 0}$$

(5)

In order to approximate these by equations that can be integrated in x, some modeling assumptions must be made. Thus the mass flux ρu must be an assigned function of x. The simplest choice is

$$\rho u = \text{constant}$$

although later we will suggest a different choice as an essential ingredient of the combustion field.

A more difficult challenge is provided by the Laplacian, since this includes not only diffusion along the axis, but off-axis diffusion associated with the curvature. Now a key parameter is the Reynolds number

$$\text{Re} = \rho_f \frac{Ua}{\mu_f} \tag{6}$$

where U is the rise speed of the flame, a its radius (when $\sigma = 0$), Fig. 1. At very low Reynolds numbers we have the diffusion controlled situation that is the essential character of Zeldovich's model, but as Re is increased, and convection becomes more important, the picture will resemble more closely a classical premixed flame in which, ahead of the reaction zone, there is a well-defined preheat zone of thickness $\delta \sim \frac{\lambda}{MC_p}$ where M is the local mass flux through the zone. Roughly speaking Re $\sim$ a/δ so that this situation is likely to prevail for the physical flames for which Re is approximately 5-7. To model the Laplacian we just need an approximation that is likely to be reasonable within the preheat zone, a small interval on the axis of symmetry, and the choice we propose is

$$\nabla^2 \Big|_{\sigma = 0} \sim \frac{d^2}{dx^2} + \frac{2}{x} \frac{d}{dx} \tag{7}$$

where x is measured from the center of curvature of the flame, Fig. 1. The second term is the curvature (off-axis) term.

This choice is precisely correct if the isotherm and isopleths (in Y) within the preheat zone and at $\sigma = 0$ all have a common center of curvature (x = 0). In some average sense this is true, and an analogy can be drawn with the modeling of two-dimensional duct flow by a quasi-one-dimensional description. The latter can be a good approximation when changes in the cross-section of the duct occur on a scale much bigger than the duct width, and (2) will be a good approximation within the preheat zone when δ is large compared to a. In addition it is accurate in the limit Re $\rightarrow$ 0, for then equations (5) and (1) are identical. This suggests that it is a plausible approximation at the modest Reynolds numbers realized in practice.

To complete the model, Ω is taken to be a Dirac δ-function such that

$$\lambda \left[\frac{dT}{dx}\right] = - QB\, e^{-E/2\,RT_*}$$

$$(\rho D)\left[\frac{dY}{dx}\right] = B\, e^{-E/2\,RT_*} \tag{8}$$

$$[\cdot] \equiv (\cdot)_{hot} - (\cdot)_{cold} \ ,$$

where T_* is the flame temperature, a function of Re that equals the value (3) in the limit Re → 0. Arrhenius' kinetics for large activation energy leads to such a model, but it has also proven valuable in the discussion of finite activation energy effects [4], so that there is no assumption here that $E/_{RT*}$ is large.

The location of the δ-function defines the location of the reaction zone, and this is x = -a.

Solution is straightforward for any choice of the function ρu, but we will write down results only for ρu = constant. When this constant is zero we recover Zeldovich's results with

$$a = \frac{(\rho D)Y_f}{B}\, e^{E/2\,RT_*} \tag{9}$$

where T_* is given by (3). For non-zero Reynolds numbers

$$T_* - T_f = \frac{QY_f}{C_p^L}\, \frac{\int_{-\infty}^{-1} \frac{dx}{x^2}\, e^{Pe(x+1)}}{\int_{-\infty}^{-1} \frac{dx}{x^2}\, e^{LPe(x+1)}} \ , \tag{10}$$

$$a = \frac{(\rho D)Y_f}{B}\, e^{E/2\,RT_*}\, \frac{1}{\int_{-\infty}^{-1} \frac{dx}{x^2}\, e^{LPe(x+1)}} \ ,$$

where Pe = PrRe is the Peclet number. In the limit Pe → ∞, equations (10) simplify to

$$T_* = T_a \ , \tag{11}$$

the adiabatic flame temperature, and

$$a = \frac{(\rho D)\, Y_f}{B}\, LPe\, e^{E/2\,RT_a} \ . \tag{12}$$

Since Pe is proportional to a, this last formula determines the adiabatic flame speed U_a.

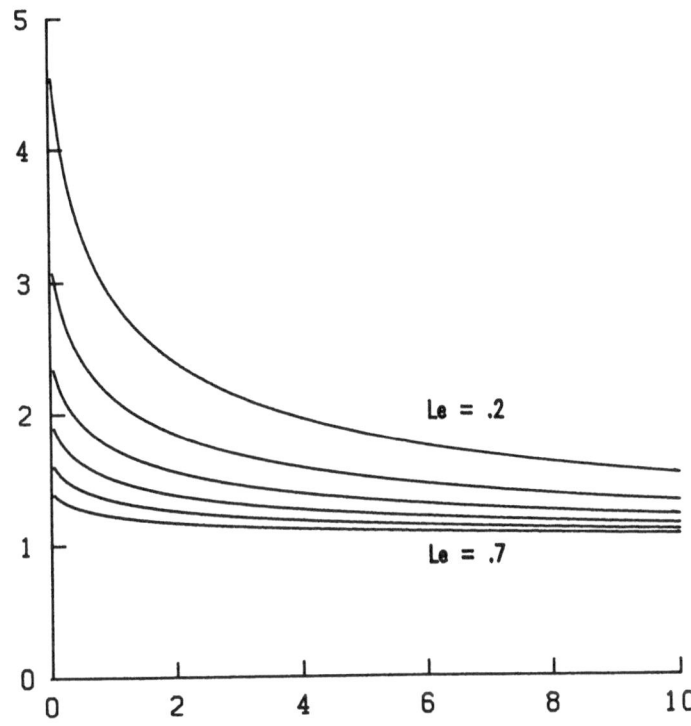

Fig. 2. Variations of $(T_* - T_f) C_p/Y_f Q$ with Pe for L = 0.2 (0.1) 0.7.

Corrections to the limiting results (11) and (12) for large but finite Pe are

$$T_* \sim T_a + \frac{2QY_f}{C_p \, Pe} \left(\frac{1}{L} - 1\right) + \ldots,$$

$$\frac{U}{U_a} \sim 1 + \frac{1}{Pe} \left[\frac{EC_p}{RQ} Y_f \left(\frac{Q}{C_p T_a}\right)^2 \left(\frac{1}{L} - 1\right) - \frac{2}{L}\right] + \ldots. \tag{13}$$

For consistency the quantity in square brackets is necessarily positive, for this is precisely the condition needed for the plane flame to exhibit cellular instabilities, the seminal mechanism for generating flame bubbles. Rewritten, this is

$$L < 1 - \frac{2}{Y_f} \cdot \frac{RQ}{EC_p} \left(\frac{C_p T_a}{Q}\right)^2 \qquad (14)$$

and (13b) is the familiar weak stretch result [5]. Here the stretch arises from the curvature, weak because Re is large.

Results for general Pe can be determined by numerical evaluation of the integrals in (10). In Fig. 2, the temperature $\frac{(T_* - T_f)C_p}{Y_f Q}$ is plotted as a function of Pe for different values of L. This function ranges from 1/L when Pe = 0, to 1 when Pe → ∞. Note that at the Reynolds numbers of practical interest there is still significant enhancement of T_* over T_a, but it is much smaller than the limit result when Re = 0.

Important Characteristics of the Flow-Field

The results of Fig. 2 are calculated for constant ρu. Although this is likely to be a reasonable approximation in an average sense in the neighborhood of the flame sheet, in a larger sense ρu will vary significantly with x. In this section we will explain why this is so, and why this might be relevant to the stability dilemma identified earlier.

The flame is a source of heat and this is swept into the wake by convection. Some estimates characterizing the wake solution are possible in the limit x → ∞. Thus the cross-sectional area ~ x so that since the flux of heat responsible for the incremental temperature in the wake is constant, this increment $\delta T_w \sim \frac{1}{x}$. From the equation of state it follows that $\delta \rho_w \sim \frac{1}{x}$, and the total (integrated) buoyancy force on the wake fluid ~ x^2. This force must be balanced by a momentum flux deficit so that $\delta(\rho u^2)_w \sim 1$. Thus there is an O(1) velocity deficit in the wake which persists as x → ∞.

A similarity solution can be constructed that provides a precise description of this deficit [6],[7]. In the notation of [7] the similarity equations have the form

$$\eta f''' + (1 + f) f'' - t = 0$$

$$\qquad (15)$$

$$\eta t' + Pr\, ft = 0$$

where $u/U = f'(\eta)$ is the x-velocity, $t = Fr^2(T/T_f - 1)x$ is a temperature function

with $Fr = U / \sqrt{ga}$ the Froude number, and $\eta = \frac{Re\,\sigma^2}{4ax}$.

Boundary conditions are

$$\eta = 0 \quad f = 0 \ , \quad f'' - t = 0, \quad f' = \alpha$$

$$\eta \to \infty \quad f' \to 1 \tag{16}$$

where different choices of α correspond to different amounts of heat flux $\dot{Q}$ from the flame, since

$$\frac{\dot{Q}}{2\pi\lambda aT_f} = \dot{q} = \frac{Pe}{Ua^2} \int_0^\infty d\sigma \ \sigma \ u \left(\frac{T}{T_f} - 1\right) = 2PrFr^2 \int_0^\infty f't \ d\eta \ . \tag{17}$$

When $Pr = 1$, t and f are related by the equation

$$t - f'' = 0 \tag{18}$$

If this is multiplied by f' and integrated from $\eta = 0$ to ∞ we find

$$f'(0) = \pm \ \left[1 - \dot{q}/Fr^2\right]^{1/2} \tag{19}$$

where the (+) sign, corresponding to a solution without reverse flow, is the most sensible choice on physical grounds. This result shows that for a given heat release there is a minimum Froude number (i.e. rise speed). If the flame rises too slowly a plume would form above it, which makes little physical sense.

The condition

$$Fr \geq (Fr) \ min., \tag{20}$$

where (Fr) min. depends on the heat release, is true for all Pr, but only for Pr = 1 can an explicit formula for (Fr) min be given. When Fr = (Fr) min. the center-line velocity in the wake is zero.

Other aspects of the flow field identified in [7] will be reported in [8], together with stability calculations for the cap model, but here we will be content with identifying a possible stability mechanism suggested by the wake solution.

How does ρu evaluated on the center-line vary with x? If we assume a monotonic variation (this is suggested by the results of [7]) then Fig. 3 will be valid with ρu decreasing significantly as we pass from the undisturbed flow ahead of the flame, to the wake solution far behind it.

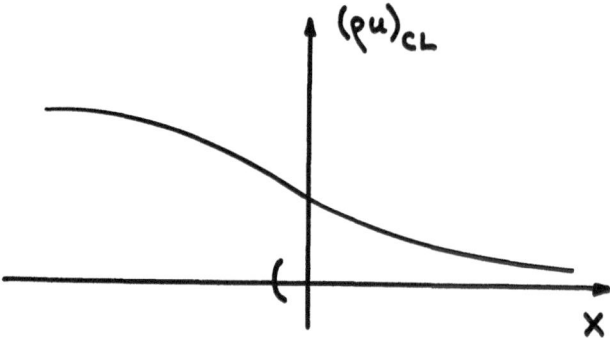

Fig. 3. Variations of the mass flux on the center-line.

The gradient, if large enough, has an obvious stabilizing effect. Displacement of the flame to the left (right) puts it into a region of increased (decreased) ρu, which will tend to drive it back to the right (left).

Acknowledgment

This work was supported by the Air Force Office of Scientific Research.

References

[1]   B. Lewis and G. von Elbe 'Combustion Flames and Explosions of Gases', Academic Press, NY, 1961, p. 314.

[2]   Ya. B. Zeldovich, G. I. Barenblatt, V. B. Librovich, and G. M. Makhviladze, 'The Mathematical Theory of Combustion and Explosions', Consultants Bureau, NY, 1985, p. 327.

[3]   Champion, M., Deshaies, B., Joulin, G., and Kinochita, K., <u>Combustion and Flame</u>, <u>65</u>, pp. 319-337, 1986.

[4]   J. D. Buckmaster and G.S.S. Ludford, 'Lectures on Mathematical Combustion', SIAM, Philadelphia, PA, 1983, p. 73.

[5]   J. D. Buckmaster and G.S.S. Ludford (ibid) p. 42.

[6]   J. Buckmaster and S. Weeratunga, <u>Combustion Science and Technology</u>, <u>35</u>, pp. 287-296, 1984.

[7]   S. Weeratunga, 'Numerical Solution of Buoyancy Induced Motion of a Hydrogen Flame Bubble', Ph.D. Thesis, University of Illinois at Urbana-Champaign, Urbana, IL, 1986.

[8]   S. Weeratunga, J. Buckmaster and R. E. Johnson, 'The Flow Field and Stability of Hydrogen Flame Bubbles', paper in preparation.

CHARACTERIZATION OF TURBULENT PREMIXED FLAME STRUCTURE
FOR MATHEMATICAL MODELING OF COMBUSTION

Akira Yoshida
Department of Mechanical Engineering, Tokyo Denki University
2-2 Kanda-Nishikicho, Chiyoda-ku, Tokyo 101, Japan

1. Introduction

Historically, the studies of turbulent premixed flames have been addressed to better understanding of turbulent combustion phenomena on the engineering standpoint. The emphasis has been placed on the measurements of the parameter which is important for designing the practical combustion devices. As a result, the data which is useful practically, for example, the turbulent burning velocities, have been accumulated extensively. However, the underlying physics of turbulent combustion was rather ambiguous. Such a situation was produced by facts that the practical requirements of the times were always urgent and that the turbulent combustion was a too complicated phenomenon to understand basically. Therefore, the mathematics could not play an important role in this field.
   Recently, the development of the new experimental technique makes it possible to make clear the essential features of turbulent premixed flames[1-3]. These should be useful for the basic scientist to develop a new model. It would be expected, as in other fields of basic sciences, that the applied mathematicians could provide a new aspect of turbulent combustion which should be surprising for phenomenalists.

2. Identification of Turbulent Combustion Regimes

The turbulent combustion includes many physical processes diversely. These are characterized by the turbulence Reynolds number and the Damkohler number which is the ratio of characteristic flow time to characteristic chemical time. Various regimes of turbulent premixed combustion can be identified depending on what the controlling process is[4,5]. In general, two important regimes are identified, namely, the wrinkled laminar flame and the distributed reaction zone. Figure 1 shows two extreme cases. When the turbulence Reynolds number is relatively small and the Damkohler number is large, the combustion reactions are rapid compared with the turbulence processes, so that the combustion tends to occur in a thin laminar flame which is wrinkled by unburned mixture turbulence. On the other hand, when the Damkohler number is small, the turbulent mixing is rapid compared with the chemistry, so that the combustion reactions are distributed throughout the volume occupied by the turbulent flame. The latter is difficult to realize in the laboratory, because such an extremely strong turbulence can not be generated easily. In the practical situation, the turbulent

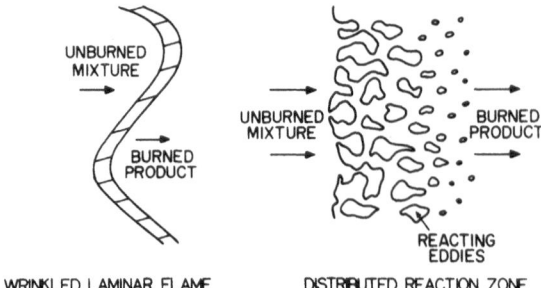

WRINKLED LAMINAR FLAME
(D_a = LARGE)

DISTRIBUTED REACTION ZONE
(D_a = SMALL)

Figure. 1. Wrinkled laminar flame and distributed reaction zone.

premixed flame is composed of the wrinkled laminar flame. Therefore, the emphasis of the present study is placed on the mechanism of the laminar flame wrinkling with the aim of the mathematical modeling of turbulent premixed flame.

3. Wrinkled Laminar Flame

Figure 2 shows a typical schlieren photograph of the wrinkled laminar flame stabilized on a Bunsen-type burner. The instantaneous structure of the turbulent flame zone consists of the continuous wavy laminar flame front. On the basis of the physical standpoint, the origin and the mechanism of flame wrinkling would be interesting. The origin can be divided into two categories, namely the hydrodynamic disturbances and laminar flame instabilities. From the early stage of combustion studies, it has been said that the laminar flame corrugates following the irregularities of the unburned mixture velocity. This hydro-dynamic reasoning is based on the Damkohler's model[6] and has been approved for a long time. However, it is doubtful that the mixture turbulence is only a determining factor for flame wrinkling, because the wrinkled laminar flame is produced even under a certain laminar

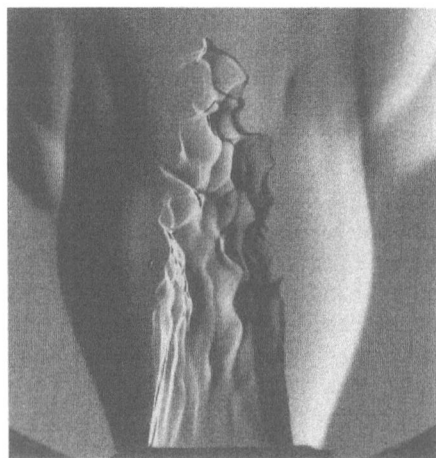

Figure 2. Wrinkled laminar flame stabilized on Bunsen type burner.

condition.

The most common flame instability results from the diffusional-thermal mechanism. This is usually called as cellular instability. This type of instability is observed only in the rich flame for the common hydrocarbon fuels. It should be noted that the wrinkled laminar flame structure appears even in the lean hydrocarbon flame as well as in the rich flame. On the other hand, the cellular structure of hydrodynamic origin can be also observed, for example, due to the Landau type instability or the Kelvin-Helmholtz type instability. Considering that the wrinkles appear in the lean flame as well as in the rich flame, the diffusional-thermal mechanism seems not so dominant for the flame wrinkling. In the present study, it will be shown that the hydrodynamic instability affects the flame wrinkling in the turbulent premixed flame.

4. Measurements of Wrinkle Size

Figure 3 shows a schematic of the time-averaged shape of the wrinkled laminar flame based on the schlieren photographs. The burned gas is separated from the unburned mixture by a thin laminar flame front of which thickness is negligible in comparison to the scale of wrinkles. For the wrinkle size measurements, the temperature is a most convenient characteristic which discriminates the burned gas from unburned mixture. If the temperature probe is placed in the wrinkled laminar flame zone, it is exposed to burned gas and unburned mixture alternately as shown in Fig. 3. Therefore, the temperature signal should be a sequence of pulses, high and low levels corresponding to burned gas and unburned mixture.

Recently, newly-developed Laser diagnostics have been widely proposed for the temperature measurements in the combustion zone. However, most of them are very expensive and difficult to operate for combustion researchers. In addition, in the reduction of Laser signal to temperature, there still remain uncertain assumptions. Here it should be noted that we only need to discriminate the burned gas from unburned mixture and that we need not know the absolute value of the gas temperature. In the relatively low frequency range (up to 5 kHz), the compensated thermocouple can afford valuable information on the

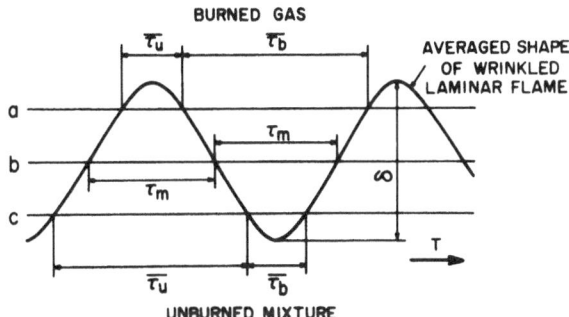

Figure. 3. Schematic of wrinkled laminar flame.

size of wrinkles[7,8]. In spite of many criticisms on the use of
thermocouple, it seems most suitable for the discrimination of burned
gas from unburned mixture in the turbulent premixed flame zone.

In the present study, a thermocouple of which wire diameter is 50
μm was used. The wire is coated by thin silica film to eliminate the
catalytic effect of platinum. Figure 4 shows the data acquisition and
processing system. An electromotive force is amplified and linearized,
and then fed to analogue or digital filter and compensation system[9].
The time scales of burned gas and unburned mixture are measured statistically, which are converted into the size of wrinkles by using the
convection velocity of wrinkles[10].

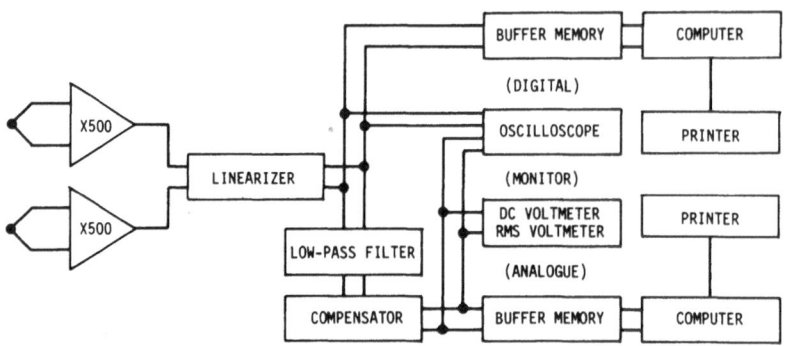

Figure 4. Data acquisition and processing system.

Figure 5 shows the typical probability density functions of
temperature signal obtained in a wrinkled laminar flame stabilized on
a Bunsen type burner. Each PDF shows a characteristic bi-modal distribution. The lower and higher peaks correspond to the unburned mixture
and the burned gas respectively. From the unburned to burned side, the
probability of unburned mixture decreases, whereas that of burned gas
increases. The probability of the intermediate temperature which corresponds to the intervening wrinkled laminar flame is negligible
properly.

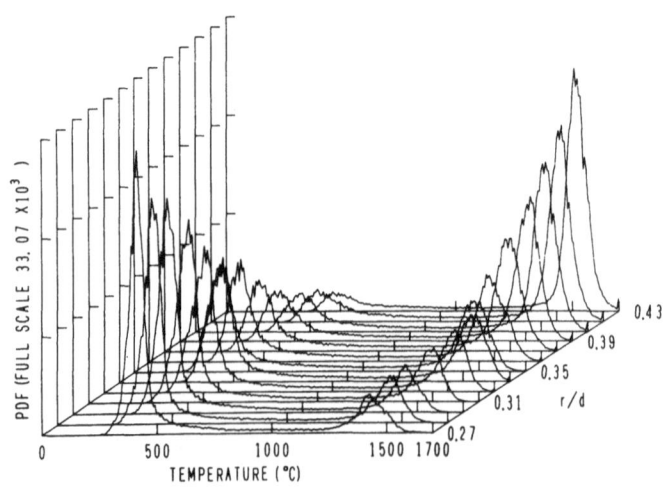

Figure 5. Probability
density functions in
a wrinkled laminar
flame.

5. Hydrodynamic Instability of Laminar Flame

It is well known that the two dimensional flame stabilized by a flame holder shows wrinkles same as in the turbulent premixed flames. A typical example is shown in Fig. 6. The unburned mixture comes from the bottom and the wrinkled laminar flame attaches to the rod. For

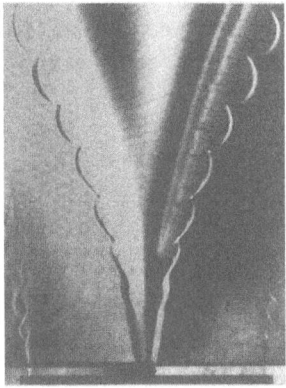

Figure 6. Laminar flame instability.

this case, the burner is unconfined and the burned gas can expand laterally. It should be noted that the unburned gas which enters the wrinkled laminar flame is absolutely laminar except for the vicinity of the flame holder. Behind the flame holder, an eddy region is formed which retards the stream locally. Thus, a shearing stress is developed owing to the high velocity gradient between the unburned mixture flow and the recirculating burned gas in the eddy. Immediately downstream of the flame holder, small indentations are triggered by this strong shearing stress and develop into regular symmetrical wrinkles in the downstream. The size of wrinkles is constant along the flame zone, whereas the amplitude increases in the downstream where the laminar stream of the unburned mixture enters the flame zone. Therefore, the amplification should be caused by the laminar flame instability.

The dependency of the size of wrinkles on the equivalence ratio is shown in Fig. 7. The wrinkle size is independent of the flame holder diameter and depends only slightly on the equivalence ratio. It

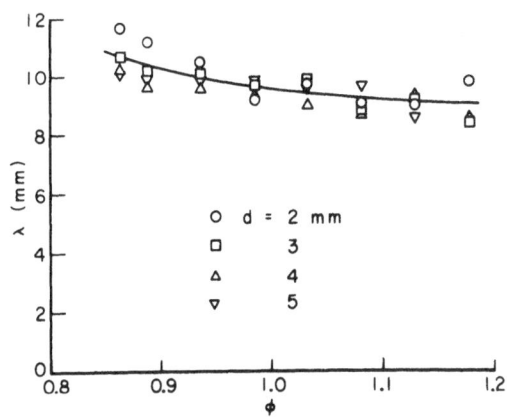

Figure 7. Variation of wrinkle size with equivalence ratio and diameter of flame holder.

is important that the size of wrinkles produced by the laminar flame
instability is about 10 mm, because this size is a standard for further discussion of wrinkle size in turbulent premixed flames.

When the burner is confined, the lateral expansion of the burned
gas is restricted. Figure 8 shows s typical shadowgraph of the wrinkled laminar flame stabilized by a rod in the constant area duct and a
schematic of the associated flowfield. The unburned mixture approaches
the flame holder with a uniform velocity. Behind the flame holder, an
eddy region is formed as the unconfined case. The unburned mixture

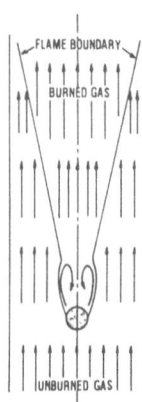

Figure 8. Shadowgraph and flowfield of
wrinkled laminar flame in confined duct.

flow attains a somewhat increased velocity due to decrease of stream
cross section by the obstacle. Therefore, the high velocity gradient
is generated again in this region. Although the burned gas expands
both laterally and axially, the axial expansion becomes more and more
predominant because the lateral expansion is restricted by the side
walls. As a result, another region with high velocity gradient is
formed in the downstream. The unburned mixture is accelerated owing to
decrease of the stream cross section by the lateral expansion of
burned gas.

Small indentations are again generated immediately downstream of
the flame holder, of which initial size is about 10 mm. In the downstream, these indentations develop into regular symmetrical wrinkles,
which in this case are stretched due to the unburned gas acceleration.
Therefore, the size of wrinkles increases in the downstream region.

Figure 9 shows typical shadowgraphs taken with changing the
unburned mixture velocity and keeping the equivalence ratio constant.
For low velocity cases, the initial indentations of which size is
again about 10 mm are stretched in the downstream due to the unburned
gas acceleration. With increasing the unburned mixture velocity, the
degree of stretch increases and wrinkles in the downstream grows more
and more. At 4.75 m/sec, small wrinkles appear on the strongly
stretched wrinkles. Further increase of the unburned mixture velocity
does not change significantly the size of these small wrinkles.

For low velocity cases, the wrinkles are stretched in the downstream, so the mean sizes of large and small wrinkles are defined here
as averages from the flame holder to 100 mm downstream. The variation

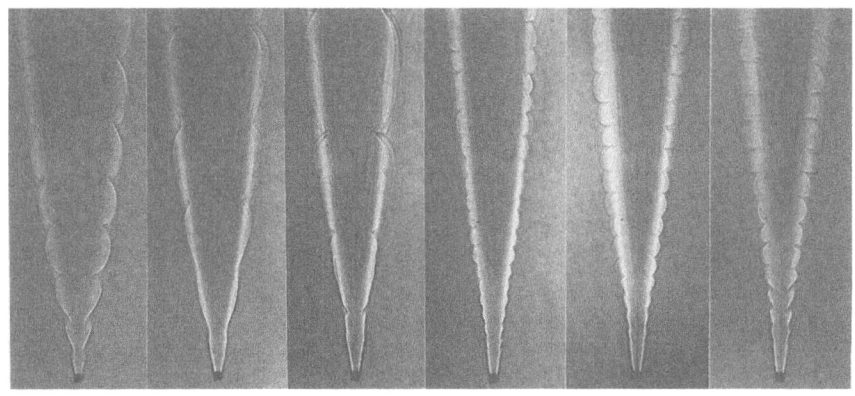

(a)   (b)   (c)   (d)   (e)   (f)

Figure 9. Variation of wrinkle size with mixture velocity.
(a)2.19 m/sec, (b)3.29 m/sec, (c)4.75 m/sec, (d)5.48 m/sec,
(e)6.57 m/sec, (f)7.30 m/sec.

of mean sizes with unburned mixture velocity is shown in Fig. 10. The
size of initial indentations is always about 10 mm, but the degree of
stretch in the downstream increases with unburned mixture velocity.
Therefore, the mean size of large wrinkles defined here increases with
the velocity. At about 4 m/sec, small wrinkles appear (half-closed
square symbols are exceptional, and will be discussed later). The mean
size of small wrinkles increases only slightly with the velocity. It
is important that the mean size of small wrinkles is about 5 mm. Thus,
two types of instability could occur on the laminar flame front stabilized by a flame holder.

6. Size of Wrinkles in Turbulent Premixed Flames

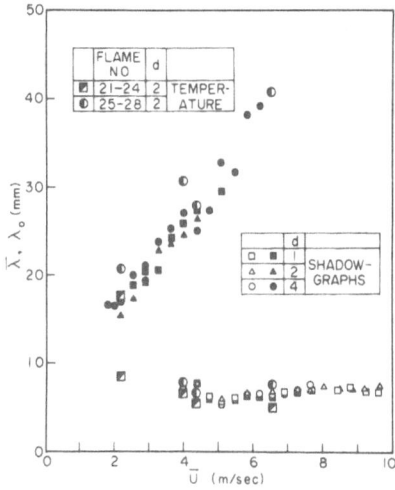

Figure 10. Size of wrinkles generated by
by laminar flame instability.

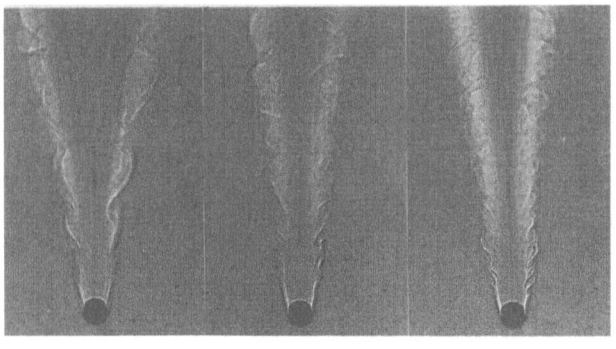

(a)   (b)   (c)

Figure 11. Shadowgraphs of turbulent premixed flames stabilized by
a flame holder:(a)2.19 m/sec, (b)4.36 m/sec, (c)6.54 m/sec.

So far, we have considered the size of wrinkles produced by the laminar flame instability. Now let's go back again to the sizes of wrinkles in the turbulent premixed flames. When the turbulence with the
relative intensity of 10 % is given to the unburned mixture in the
confined duct, the instantaneous flame shape changes as shown in Fig.
11. The regular symmetrical wrinkles are broken up into three dimensional wrinkles. It is difficult to measure the size of wrinkles from
these shadowgraphs. Therefore, the wrinkle size should be estimated
statistically by the analysis of the temperature signal.
      Figure 12 shows the probability density functions of time scales
of unburned mixture (before conversion into the size). For the laminar
flame instability, a single (for low velocity case) or two (for high
velocity case) eminent time scales are observed, because the laminar
flame instability produces only one size or two sizes of wrinkles
depending on the mixture velocity. However, in the turbulent premixed
flames, the time scale distributions are rather broad, ranging from
0.1 msec to 10 msec. The turbulence characteristics of the unburned
mixture apparently affect the profiles of the probability density
functions of the wrinkle size. Here it should be noted that the eminent time scale exists in each distribution. This means that there

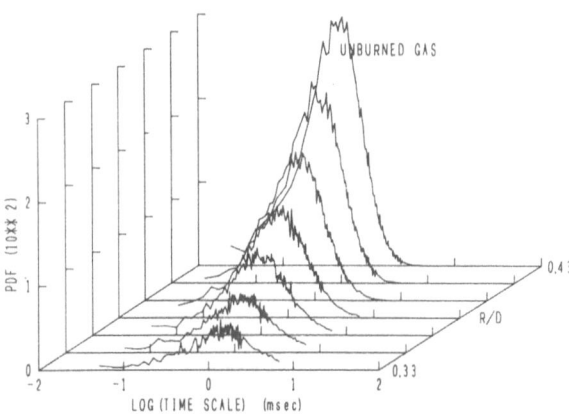

Figure 12. Probability
density functions of
time scales of unburned
mixture.

exists a wrinkle size which is most probable in the turbulent premixed flame zone.

The half-closed square symbols in Fig. 10 show the eminent size of wrinkles when the unburned mixture is turbulent. For U = 2.19 m/sec, two eminent sizes are observed. It is clear that the larger size corresponds to the first type instability of laminar flame. Therefore, even if the turbulence is given to the unburned gas, large wrinkles generated by the laminar flame instability remain in the turbulent premixed flames. The smaller size corresponds to the second type instability of laminar flame. This second type instability is caused by the local shearing stress in the turbulent mixture flow. With increasing the mixture velocity to 4 m/sec, wrinkles with larger eminent size disappears. This fact suggests that the large wrinkles are broken up into the smaller wrinkles by the mixture turbulence. The eminent size coincides with the size of wrinkles generated by the second type instability of the laminar flame. Of course, the wrinkle size distributes broadly when the unburned mixture is turbulent. However, the most probable size coincides with that of second type instability of laminar flame.

Figure 13 shows the mean and eminent sizes of wrinkles in the Bunsen-type turbulent premixed flames under many unburned mixture turbulence conditions. λ_f and λ_0 are the mean and eminent sizes of wrinkles and τ_a is the lapse of time when a wrinkle moves from the

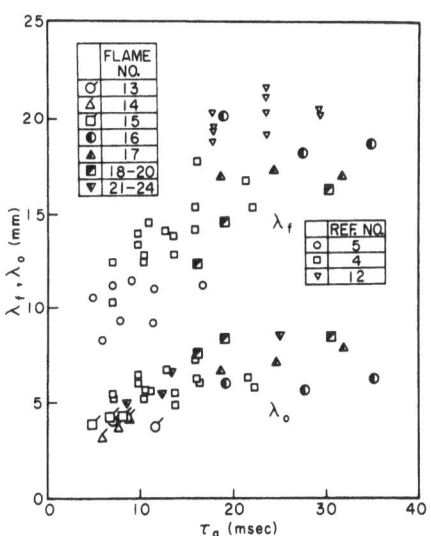

Figure 13. Mean and eminent sizes of wrinkles in turbulent premixed flames.

burner port to the measuring point along the flame blush. The indentations are first recognized at τ_a = 4 msec and the minimum size at this point is about 10 mm. It should be noted that the size of wrinkles generated by the first type instability of laminar flame is about 10 mm. This coincidence suggests that the first indentations of the turbulent premixed burner flame is caused by the laminar flame instability which should be triggered by the unburned mixture turbulence. Of course, the effect of the turbulence characteristics is not negligible in the process of downstream development of wrinkles. Wrinkles smaller and lager than the mean size are also generated by the mixture turbulence, but the small wrinkles collapse rapidly into large wrinkles in the downstream. Therefore, the mean size increases with τ_a.

The eminent size is independent of the burner geometry and turbulence characteristics of the unburned mixture. The eminent size of wrinkles measured in the turbulent premixed burner flame is about 5 mm, which slightly increases with τ_a. This eminent size coincides with the wrinkle size of second type instability of the laminar flame. Therefore, it can be concluded that the laminar flame instability plays an important roll in the flame wrinkling in the turbulent premixed flames.

7. Conclusions

Measurements of wrinkle size are made extensively in laminar and turbulent premixed flames to find out the origin of flame wrinkling in turbulent premixed flames, and conclusions are summarized as follows:
(1) Two types of instability are observed on the laminar flame front stabilized by a flame holder. These instabilities are caused by the strong shearing stress developed by the high velocity gradient between the unburned mixture and burned gas.
(2) The laminar flame instability plays an predominant role in flame wrinkling and the unburned mixture turbulence triggers the two types of laminar flame instability in the turbulent premixed flames.
(3) It is found qualitatively that the mixture turbulence broadens the wrinkle size distributions.

References

1 Yoshida, A. and Tsuji, H.:Seventeenth Symposium (International) on
   Combustion (The Combustion Institute, 1978), pp.945-956.
2 Yoshida, A. and Gunther, R.:Combust. Flame 38, 249-258(1980).
3 Yoshida, A.:Eighteenth Symposium (International) on Combustion
   (The Combustion Institute, 1980), pp.931-939.
4 Williams, F. A.:Combustion Theory, 2nd ed.(The Benjamin/Cummings,
   1985).
5 Bray, K. N. C.:'Turbulent Flows with Premixed Reactants,' Turbulent
   Reacting Flows, Topics in Applied Physics, Vol. 44, ed. by P. A.
   Libby and F. A. Williams (Springer, 1980).
6 Damkohler, G.:Z. Elektrochem. 46, 601-626(1940).
7 Yoshida, A. and Tsuji, H.:Nineteenth Symposium (International) on
   Combustion (The Combustion Institute, 1982), pp.403-411.
8 Yoshida, A. and Tsuji, H.:Twentieth Symposium (International) on
   Combustion (The Combustion Institute, 1984), pp.445-451.
9 Yoshida, A. and Nakamura, S:Combust. Sci. Tech., submitted (1986).
10 Yoshida, A.:Twenty-first Symposium (International) on Combustion
   (The Combustion Institute, 1986), in press.

THE STABILITY OF WEAKLY STRETCHED FLAMES

M. Matalon
Engineering Sciences and Applied Mathematics
The Technological Institute
Northwestern University
Evanston, IL 60208

ABSTRACT

In this study, the stability of plane stretched flames, more specifically plane flames in straining fields, has been examined. It is shown that flame stretch stabilizes long wavelength disturbances and so can suppress, in this regime the hydrodynamic instability. If in addition, the mixture is deficient in the reactant which is also the weakly diffusing component, and hence the Lewis number is greater than unity, thermal effects will stabilize the short wavelength disturbances. Thus sufficiently strong stretch can render a flame absolutely stable. The instability, which first appears by reducing the strain rate from the critical value, is in the form of longitudinal cells with ridges in the direction of stretch. By reducing the strain rate further a cellular structure will probably emerge.

INTRODUCTION

The simplest model of a steadily propagating, plane, adiabatic flame treats the whole flame as a discontinuity characterized by two parameters: the temperature rise across the flame, or the flame temperature T_b^0, and the propagation speed S_f^0, both being characteristics of a combustible mixture. Whereas the flame temperature is calculated from an overall energy balance (Hugoniot curve and Rayleigh line type analysis), the internal structure of the flame must be considered for the determination of the propagation speed. Indeed, this configuration must be stable for combustion to be actually maintained in the form of a propagating plane wave.

The earliest stability analyses were carried out by Darrieus [1] and by Landau [2]. They treated the flame as a hydrodynamic discontinuity thus considering disturbances of wavelength L much larger than the flame thickness L_D, where $L_D = D_{th}/S_f^0$ and D_{th} is the thermal diffusivity of the mixture. Furthermore, they assumed that the perturbed flame maintains the same temperature T_b^0 and propagates at the same speed S_f^0 as the unperturbed flame. Their analyses reveal that the growth rate ω of a disturbance is given by

$$\omega = \omega_0(\sigma)S_f^0 k \quad , \qquad , \ \omega_0 = \frac{\sqrt{\sigma^3+\sigma^2-\sigma} - \sigma}{\sigma+1} \qquad (1)$$

where $k = 2\pi L^{-1}$ is the wavenumber and σ is the density ratio of the unburnt gas to that of the burnt gas (also equal to the temperature ratio of the burnt gas to that of the unburnt gas), i.e.

$$\sigma = \rho_u^0/\rho_b^0 = T_b^0/T_u^0 \quad .$$

Since for all $\sigma > 1$ the growth rate ω is positive, they concluded that plane flames are unstable to disturbances of all wavenumbers. This hydrodynamic instability, therefore, does not permit the existence of plane flames.

According to equ. (1), short wave disturbances grow faster than long-wave disturbances. But for short waves, the Darrieus-Landau results are inadequate because diffusive and thermal effects within the flame may cause the flame speed and temperature to depend appreciably on the flame shape and on the local flow field. Thus, later work was aimed at improving the Darrieus-Landau results in the short wave regime. An account of the early work of the 1950's appears in [3]. The recent, more mathematically consistent studies [4]-[12], are based on the assumption that the activation energy is large. In this limit, the chemical reaction is confined to a thin reactive-diffusive zone which can be resolved and then matched on either side to the remaining convective-diffusive field. If one also adopts the framework of slowly varying flames which is concerned with disturbances of wavelength much larger than the thickness of the flame, or more precisely $L = EL_D$ where $E \gg 1$ is the activation energy parameter measured in units of the enthalpy of the fresh mixture, the whole flame may be again treated as a hydrodynamic discontinuity and analytical expressions can be derived for the flame speed S_f and flame temperature T_b. These expressions indicate that, unlike the assumption in [1] and [2], the flame speed and temperature vary along the front and depend on the flame shape and on the local flow field. These variations occur on the longer length scale L. If one instead considers disturbances of shorter wavelength which are comparable to the flame thickness, i.e. $L \sim L_D$, the flame ceases to be slowly varying and the problem is analytically intractable. Results must then be obtained numerically.

The conclusion drawn from these various studies is that diffusive and thermal effects may often introduce stabilizing influences so that, the growth rate, which behaves as in (1) for large wavelength, achieves a maximum value and decreases thereafter with decreasing L becoming negative at a critical wavelength L_c. Although in such circumstances the flame is stable to disturbances of wavelength smaller than L_c, it is unstable to disturbance larger than L_c. This means that plane flames could only be observed if the hydrodynamic instability can be suppressed for large wavelength disturbances. In this work we examine the effect of flame stretch on the hydrodynamic instability and whether a sufficiently large stretch can render the flame absolutely stable.

REVIEW OF STABILITY ANALYSES - PLANE FLAMES

Slowly Varying Flames. As a result of the large activation energy assumption, the flame temperature deviates only slightly from the adiabatic flame temperature, i.e. $T_b - T_b^0 + O(E^{-1})$. The flame speed, however, may vary appreciably depending on the relative rates of heat and mass diffusivities and on the local flow field and flame shape. The equation for the flame speed (see [4],[5]) is

$$\left(\frac{S_f}{S_f^0}\right)^2 \ln\left(\frac{S_f}{S_f^0}\right) = - \cdot \frac{L_D}{S_f^0} \frac{Eb}{2} \left[K - \frac{1}{S_f} \frac{DS_f}{Dt}\right] \tag{2}$$

where K is the flame stretch and D/Dt the convective derivative along the flame front. Note that, along the front, the flame speed may differ from the adiabatic flame speed S_f^0 by an $O(1)$ amount, because spatial and temporal variations occur on the slow time scale corresponding to L/S_f^0. The parameter b is given by

$$b = \frac{\sigma-1}{\sigma^2} \int_0^1 \frac{1-\xi^{Le-1}}{1+(\sigma-1)\xi} d\xi \quad , \tag{3}$$

where $Le = D_{th}/D_m$ is the Lewis number; D_{th} and D_m are the thermal and molecular diffusivities respectively. When $Le \to 1$, $b \to 0$; otherwise b is positive/negative for Le greater/less than unity, respectively.

If a stability analysis similar to [1] and [2] is now carried out the following dispersion relation for the growth-rate ω is obtained [4],[5]

$$(\sigma+1)[\beta\alpha^3+(\beta-1)\alpha^2] - \sigma[2+(3\sigma-1)\beta]\alpha + \sigma[\sigma-1-(3\sigma-1)\beta] = 0 \tag{4}$$

where $\alpha = \omega/kS_f^0$ and $\beta = bLk/2$. It can be easily verified that (4) posesses three real roots one of which is always negative and therefore does not correspond to an instability. The two possibly unstable modes are displayed in Figure 1 where, based on the numerical solution of (4), the growth rate ω has been plotted as a function of the wave number k. Mode (a), which reduces to the hydrodynamic instability (1) when Le = 1 (or b = 0), indicates that diffusion stabilizes short waves if Le > 1 but not otherwise. Mode (b), which appears to involve very high frequencies, indicates an absolute instability only if Le > 1. By expanding the solutions of (4) in power series of k approximate expressions for Ω^0 and Ω^1, the roots corresponding to modes (a) and (b) respectively, can be obtained as follows:

$$\Omega^0 = \omega_0(\sigma)S_f^0 k - \frac{Eb}{2}\omega_1(\sigma)D_{th}k^2 + \ldots \quad , \qquad \omega_1 = \frac{\sigma(\sigma+\omega_0)(1+\omega_0)}{\sigma+(1+\sigma)\omega_0} \tag{5}$$

$$\Omega^1 = \frac{2(S_f^0)^2}{EbD_{th}} + \frac{\sigma-1}{\sigma+1}S_f^0 k + Eb\frac{\sigma(\sigma^2+1)}{(\sigma+1)^2}D_{th}k^2 + \ldots \quad . \tag{6}$$

The conclusion is again, that plane flames are unconditionally unstable.

Near Equidiffusional Slowly Varying Flames. Equation (2) indicates that, as Le → 1 (or b → 0), the flame speed $S_f \to S_f^0$ and remains constant along the front. However, the analysis leading to (2) breaks down as Le $- 1 = 0(E^{-1})$. The variations of the flame speed along the front are also $0(E^{-1})$ and can be obtained by carrying the previous analysis to higher orders. One finds [8] that

$$S_f = S_f^0 [1 - \frac{L_D}{S_f^0} (\frac{E\hat{b}}{2} + \frac{\sigma ln\sigma}{\sigma-1})K] \quad , \tag{7}$$

with $\hat{b}$ being actually the leading term expansion of b, namely

$$\hat{b} = \frac{(Le-1)}{\sigma^2} \int_1^\sigma \frac{ln\xi}{\xi-1} d\xi \quad . \tag{8}$$

The variations of the flame temperature from the adiabatic flame temperature remain small, more precisely $T_b = T_b^0 + 0(E^{-2})$. Note that the expression (7) for the flame speed is, in fact, a linearization of (2) but with the coefficient multiplying the stretch K modified appropriately.

To leading order, the stability problem is exactly that treated by Darrieus and Landau so that, by including diffusive and thermal effects as a perturbation, one expects to obtain an expression for the growth rate similar to (5), i.e. corresponding to mode (a). And indeed, we find [7]-[9], that

$$\Omega^0 = \omega_0(\sigma)S_f^0 k - [\frac{Eb}{2}\hat{\omega}_1(\sigma) + \hat{\omega}_1(\sigma)] D_{th}k^2 + \ldots$$

$$\tag{9}$$

$$\hat{\omega}_1 = \frac{1}{2}\left\{\frac{\sigma(\sigma-1)^2+\sigma\ln\sigma[\sigma(\sigma-1)+2(1+\omega_0)]}{(\sigma-1)[\sigma+(1+\sigma)\omega_0]}\right\} .$$

Thus, for near equidiffusional flames, a Taylor expansion of the growth rate corresponding to mode (a) is given by (9). Note that, the qualitative behaviour of Ω^0 can be deduced correctly from (5) and, even more precisely, from the dispersion relation (4) which is not limited to $k \ll 1$. However, unlike (5) which indicates that diffusion has a stabilizing effect on short wavelength disturbances if $b > 0$, the relation (9) shows that this effect is stabilizing if $\hat{b} > -2\sigma\ln\sigma/E(\sigma-1)$. That is, the critical Lewis number Le = 1 has been refined here to a value close to, but slightly less than one. This refinement could be useful for estimating more precisely the critical parameters. Finally, it should be noted that, apparently, because of the relatively slow time scale adopted in deriving (7), the unstable mode (b) was not uncovered here. This mode is associated with high frequencies (see Fig. 1) possibly comparable to the transit time through the flame.

Near Equidiffusional Flames. When the flame cannot be considered slowly varying in the framework of large activation energy, one is forced [5] to adopt the assumption of near equidiffusion, i.e. Le - 1 = $0(E^{-1})$. The only limit which is tractable analytically is that corresponding to no thermal expansion ($\sigma = 1$). Here the hydrodynamic instability (1) is suppressed and one obtains the diffusional-thermal results [10] which indicates, that a narrow band of Lewis numbers exist near Le = 1, where the flame is absolutely stable. For Le < Le$_*$, with Le$_*$ < 1, an instability associated with the existence of cellular flames occurs. For Le > Le*, with Le* > 1, and instability associated with pulsating flames and/or flames supporting travelling waves along their fronts [11] occurs. In the presence of thermal expansion, the stability problem has been solved numerically [12] indicating that the cellular instability is present even for Le > Le* but for long wavelength disturbances only. The other instability remains practically unchanged. Thus, only short wavelength disturbances are stable in the band

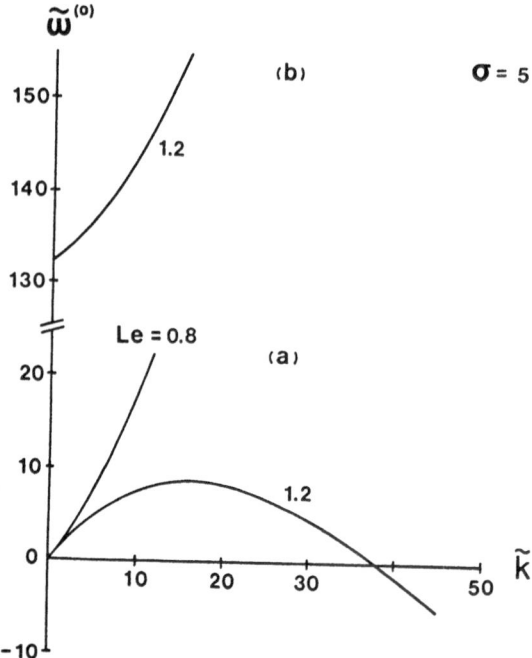

Fig. 1: The unstable modes corresponding to equation (4). Here $\tilde{\omega} = (EL_D/S_f^0)\omega$ and $\tilde{k} = (EL_D)k$. For case (a), $\omega = \Omega^0$ whereas for case (b), $\omega = \Omega^1$.

$Le_* < Le < Le^*$. Note, that this band of Lewis numbers is within $0(E^{-1})$ from $Le = 1$ and so, on an $0(1)$ scale of the Lewis number it shrinks to zero consistent with the results of the slowly varying theory. The cellular instability may be regarded as a refinement of Ω^0, i.e. mode (a) of Figure 1, because, in both cases the instability is suppressed for $Le > 1$ if $\sigma = 1$ and when $\sigma \neq 1$ the instability is restricted only to long wavelength disturbances. The instability associated with the travelling waves along the flame front may be regarded as a refinement of Ω^1, i.e. mode (b) in Figure 1, because, both are practically unaffected by the hydrodynamic instability and they are absolutely suppressed for $Le < 1$. The fact that the growth rate Ω^1 in the slowly varying theory is real, with no imaginary part representing the transverse travelling waves, is apparently related to the relatively slow time scale adopted to describe the flame front configuration in this case. On this scale, the oscillatory behavior is averaged out and only the net growth is retained.

The results summarized above suggest that diffusive and thermal effects within the flame introduce a stabilizing influence when $Le > 1$, that is

produce a critical wavelength L_c such that the flame is stable to disturbances

with $L < L_c$. This means that plane flames could only be observed if the

appearance of long wavelength disturbances, such that $L > L_c$, are prevented or

their growth is suppressed. In this, and a subsequent study [13], we shall
investigate whether flame stretch could stabilize the long wavelength
disturbances thus rendering flat flames in the presence of stretch absolutely
stable. Of practical importance is also to determine how strong should a
flame be stretched to be absolutely stable. The only other stability analysis
of a stretched flame [14] is restricted to the case $\sigma - 1$, i.e. when there is

no hydrodynamic instability and therefore addresses different questions than

ours.

What model should one use? The prediction of a stability band, if long

wavelength disturbances are suppressed, for Le - 1 - $0(E^{-1})$ and the fact that
for many combustion mixtures Le is sufficiently near unity suggest that one
should adopt the assumption of near equidiffusion. However, within the
framework of slowly varying flames this assumption leads to results of limited

range because, the flame speed (7) is only a perturbation of S_f^0 and as such

provides at best the tendency that diffusional and thermal effects may have on
the flame. For example, the stability result (9) corresponds to the first two

terms in a Taylor series of the growth rate Ω^0. In the absence of estimates
of subsequent terms in the series it is only a speculation to argue that the
second term can stabilize short wavelength disturbances that would otherwise,
on account of the first dominant term, be unstable. Indeed, in this case, the
speculation turns out to be correct and so, equ. (9) valid for $k \ll 1$, was
found useful in computing critical parameters, such as the dependence of the
critical wave number on the Lewis number, by extrapolating the result to $0(1)$

values of k. This difficulty is removed if one does not restrict attention to
Lewis numbers near unity because as remarked earlier, both sides of equation
(2) for the flame speed are $0(1)$. And indeed, unlike (9), the dispersion
relation (4) is not restricted to small values of k. Futhermore, as noted
above this dispersion relation appears to contain, at least qualitatively, the
near equidiffusional result. Also, values as large as 1.82 for the Lewis
number, which holds for propane-air mixtures as reported in [15], are not
necessarily near unity and are not uncommon. Based on these arguments, it
seems preferable to adopt the slowly varying model with Le - 1 - $0(1)$. The
problem is that equation (4) predicts absolute instability because, the
unstable mode corresponding to a flame supporting travelling waves, i.e. mode
(b), is always present for Le > 1. Nevertheless, since our primary goal here
is to examine the effect of stretch on the cellular instability, we shall use
this model and concentrate only on that mode. Refinement may later be

obtained by reexamining the problem for Lewis numbers near unity and so, even though extrapolation is needed in that case in order to extend the results from small to moderate values of k, one can do that with enough confidence given the conclusions of the present study which are valid for all k. Of course, one can abandon the slowly varying framework and consider disturbances of wavelength comparable to the flame thickness, but this leads to a nontrivial numerical problem which will be addressed in the future.

THE GOVERNING EQUATIONS FOR SLOWLY VARYING FLAMES

The governing equations consist of Euler's equations

$$\rho \frac{D\vec{v}}{Dt} = -\nabla p \quad , \quad \nabla \cdot \vec{v} = 0 \tag{10}$$

subject to the jumps

$$\left[\rho (\vec{v} \cdot \vec{n} + V_f) \right] = 0$$

$$\left[p\vec{n} + \rho \vec{v} (\vec{v} \cdot \vec{n} + V_f) \right] = 0 \tag{11}$$

across the flame front and equation (2) for the flame speed S_f. Here

$\vec{v} = (u,v,w)$ is the velocity field, p is the pressure, ρ is the density and t is the time variable. If $\Phi(\vec{x},t) = 0$ is the equation describing the instantaneous position of the flame front, the unit normal $\vec{n}$ and velocity V_f are given by

$$\vec{n} = \frac{\nabla\Phi}{|\nabla\Phi|} \quad , \quad V_f = - \frac{\partial\Phi}{\partial t}/|\nabla\Phi|$$

and the flame speed, by definition, is $S_f = \vec{v} \cdot \vec{n} + V_f$ with the right hand side evaluated just ahead of the flame. Finally, the density is a piecewise constant function, to leading order in E^{-1}, with ρ_u and ρ_b its values in the unburnt/burnt sides respectively.

STABILITY ANALYSIS - STRETCHED FLAMES

The simplest circumstance of a flat flame which experiences a constant stretch is when it stands in the front stagnation point flow of a body. For moderate values of the strain rate, or stretch, the flame lies in the inviscid flow well ahead of the body. For example, for a strain rate $\epsilon \sim 50 \ s^{-1}$ the flame standoff distance is about 100 L_D, see [16], and thus can be treated as slowly varying. A full description of the flow field including the determination of the flame standoff distance was given, within this framework, in [17] for both the plane and the axisymmetric cases. Here, we consider the plane case. Furthermore, we restrict attention to a weak stretch, i.e. $\epsilon \ll S_f^0/EL_D$, where the flame recedes back upstream in the nearly uniform flow. The case of an arbitrary ϵ is treated in a sequel [13]. If the steady state solution described in [17] is specialized for small ϵ and, if a coordinate system (x,y,z) attached to the flame is considered (where x measures distances normal to the flame front and y,z distances along the flame front, the former being in the direction of the strain), the flow field is given by

$$u = f(x) , \qquad v = - yf'(x) , \qquad w = 0 \tag{12}$$

$$f(x) = \begin{cases} 1 - \epsilon(x + b/2) + \ldots , & \text{for } x < 0 \\ \sigma - \epsilon(x + \sigma b/2) + \ldots , & \text{for } x > 0 \end{cases}$$

with prime denoting differentiation. Now, the body is at a distance $1/\epsilon$ from the flame located at $x = 0$ and so, as $\epsilon \to 0$, x extends downstream all the way to infinity.

For the stability problem, we first introduce small disturbances superimposed to the steady state (12). Since the basic state varies with y, a disturbance is regarded small only when compared to the basic state at the same station of y. This suggests writing

$$u = \quad f(x) \quad + \quad U(x) \ \exp[ikz+\omega t]$$
$$v = -yf'(x) + yV(x) \ \exp[ikz+\omega t]$$
$$w = \qquad\qquad W(x) \ \exp[ikz+\omega t]$$
$$\Phi = \qquad\qquad A \ \exp[ikz+\omega t]$$

Substituting into the governing equations and then linearizing about the steady state we obtain the following problem, for the determination of the growth rate ω, correct to $O(\epsilon)$.

Equations:
$$fV' + (\omega+2\epsilon)V = 0$$
$$f(W'-ikU)' + (\omega-\epsilon)(W'-ikU) = 0$$
$$U' + V + ikW = 0$$

Jump conditions at $x = 0$:
$$[\rho U] = - \frac{\sigma-1}{\sigma}(\omega+\epsilon)A , \qquad [V] = 0,$$

$$[W] = -ik(\sigma-1)S_f A, \qquad [W'] + \omega\, S_f^{-1}[\rho W] = 2ik[U].$$

Flame speed equation at $x=0^-$:

$$\left(S_f^2(1+\ln S_f^2) - \frac{b}{2}\omega\right)\left[U - (\omega+\epsilon)A\right] = \frac{b}{2}S_f\left[k^2 S_f A - V - ikW\right],$$

where $S_f = 1-\epsilon b/2$ is the flame speed of the unperturbed flame.

To leading order in ϵ the problem is identical to that of a plane flame in a uniform flow. Thus, to leading order, ω is determined from the dispersion relation (4). For the cellular instability the growth rate is given by $\Omega^0 = \alpha_0(\sigma,\beta)S_f^0 k$ with α_0 the appropriate root of equation (4); the dependence of Ω^0 on k is shown in Figure 1. If the analysis is carried out to the next order in ϵ, the effect due to stretch appears as a correction to Ω^0 and we obtain

$$\omega = S_f^0 k\,[\alpha_0(\sigma,\beta) + \frac{\epsilon L_D}{S_f^0}\,\alpha_1(\sigma,\beta) + \ldots]. \tag{13}$$

The expression for $\alpha_1(\sigma,\beta)$ is lengthy so that we shall avoid writing it in full. The proper behavior can be seen from its expansion in power series of k, namely

$$\omega = \omega_0(\sigma)S_f^0 k - \frac{Eb}{2}\,\omega_1(\sigma)\,D_{th}k^2 - \epsilon\omega_2(\sigma) + \ldots \tag{14}$$

$$\omega_2 = \frac{2\omega_0[8\sigma^4+25\sigma^3+11\sigma^2-13\sigma+1] + \sigma(\sigma-1)(\sigma^2-4\sigma-9)(\sigma^2+2\sigma-1)}{2(\sigma-1)[\sigma^5+3\sigma^4-\sigma^3-3\sigma^2+4\sigma-(2\sigma^4+5\sigma^3-3\sigma^2-9\sigma+1)\omega_0]}$$

It can be verified that ω_2 is positive for all $\sigma > 1$, varying from 1 when $\sigma = 1$ to 0.5 when $\sigma \to \infty$, which suggests that long wavelength disturbances can be stabilized by stretch. Since short wavelength disturbances are stabilized by diffusional and thermal effects when Le > 1, the range of disturbances for which the flame is unstable in the presence of stretch will be limited to some moderate wavelength only. An examination of equation (13) with the full expression for α_1 shows that indeed this is the case (see Fig. 2). Therefore, it may be expected that a sufficiently strong stretch will suppress the hydrodynamic instability and makes the flame absolutely stable. Equation (13), which is valid for small ϵ, indicates also that tendency. For arbitrary ϵ the function f(x) is no longer linear in x so that the resulting eigenvalue problem for ω must be solved numerically. The results of that study [13] show

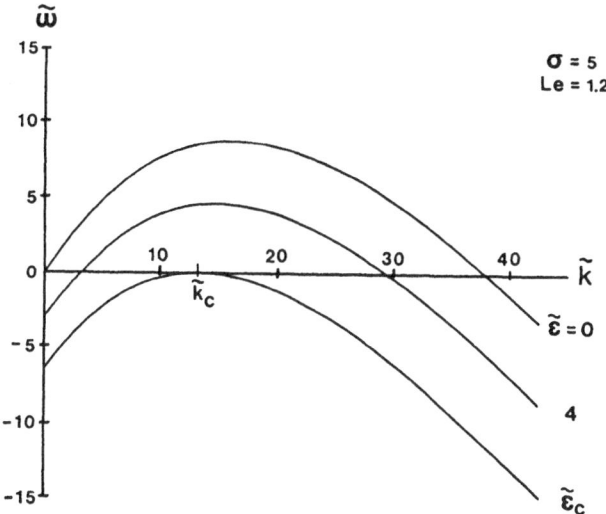

Fig. 2: The growth rate $\tilde{\omega} = (EL_D/S_f^0)\omega$ versus the wave number $\tilde{k} = (EL_D)k$ for different values of the stretch $\tilde{\epsilon} = (EL_D/S_f^0)\epsilon$. The critical values are $\tilde{\epsilon}_c = 8.52$ and $\tilde{k}_c = 13.23$.

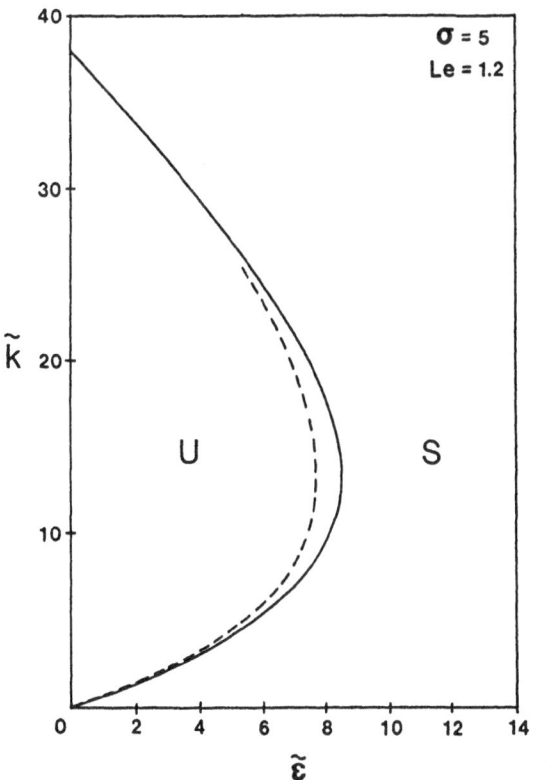

Fig. 3: Neutral stability curves. Broken curve is based on the numerical results of [13], solid curve based on the approximation (13).

that, indeed, the flame is absolutely stable (to small disturbances of the type assumed) if $\epsilon > \epsilon_c$. The neutrally stable curve $\epsilon_c(k)$ is plotted in Fig. 3 (the broken curve). We have also plotted in this figure the neutral curve based on (13) after setting ω to zero (the solid curve). Note that, despite being valid for small ϵ, equation (13) appears to be a good approximation for values of (the dimensionless) ϵ as large as 8. As ϵ decreases below ϵ_c there is a range of unstable modes which widens up and covers the range $0 < kEL_D < 2b^{-1}(\sigma-1)/(3\sigma-1)$ as $\epsilon \to 0$ consistent with (4).

In Fig. 4 the neutral curves are plotted for various values of the Lewis number. For Le < 1 the short wavelength disturbances remain unstable whereas for Le > 1 the flame is stable if $\epsilon > \epsilon_c$. The larger the Lewis number the smaller the ϵ_c needed to achieve absolute stability.

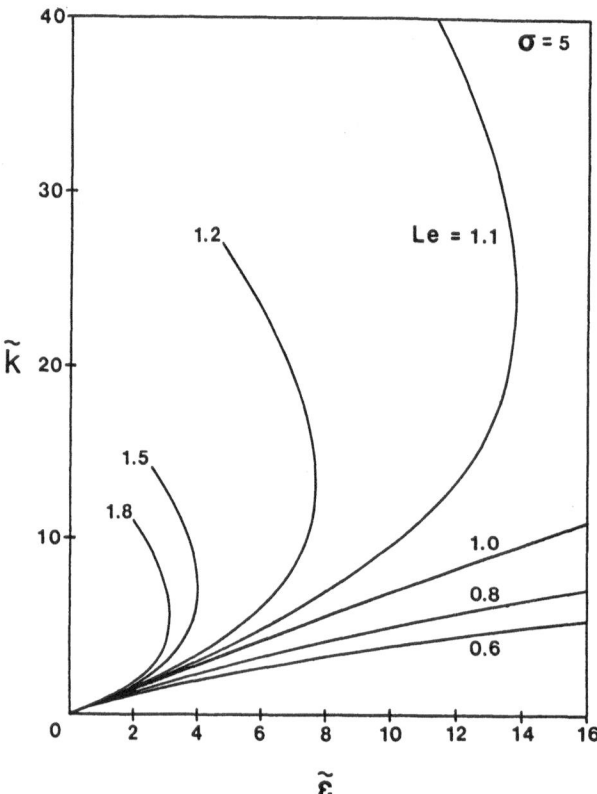

Fig. 4: Neutral stability curves for different Lewis number.

CONCLUSIONS

We have shown that the effect of flame stretch is to stabilize the long wavelength disturbances which would otherwise grow as a result of the hydrodynamic instability. If in addition the Lewis number is greater than unity, diffusional and thermal effects would stabilize the short wavelength disturbances. Therefore, a sufficiently strong stretch could render a flame absolutely stable consistent with observations. Experiments performed in an axisymmetric straining flow field [18],[19] indicate the appearance of cellular flames when the strain is very weak, flames with ridges along the radial direction (star shaped) at moderate strain rates and smooth flat flames at larger strain rates. Our analysis indicates that, when the instability is first set, it appears in the form of longitudinal cells with ridges in the direction of stretch because, the spatially periodic structure of the disturbances considered in this study is in the direction perpendicular to stretch. This pattern has been observed experimentally [20] and is in fact the two-dimensional version of the star-shaped flames mentioned above. By reducing the stretch further this configuration will probably become unstable and a cellular structure will emerge.

ACKNOWLEDGMENTS

This work has been supported in part by the National Science Foundation under Grants CBT-8521352 and DMS-8601903 and, the Department of Energy under Grant DE-FG02-87ER-25027.

REFERENCES

1. Darrieus, G., 1945, paper given at the Six International Congress of Applied Mechanics; Paris 1946.
2. Landau, L. D., 1944, Acta Physicochimica URSS 19, 77.
3. Markstein, G. H., 1964, Nonsteady Flame Propagation, AGARDograph No.75, New York: MacMillan.
4. Sivashinsky, G. I., 1976, Acta Astronautica, 3, 889.
5. Buckmaster, J. D. and Ludford, G. S. S., 1982, Theory of Laminar Flames, Cambridge University Press.
6. Clavin, P. and Williams, F. A., 1982, J. Fluid Mechanics, 116, 251.
7. Pelce, P. and Clavin, P. 1982, J. Fluid Mechanics, 124, 219.
8. Matalon, M. and Matkowsky, B. J., 1982, J. Fluid Mechanics, 124-239.
9. Frankel, M.L. and Sivashinsky, G.I. 1982, Combusion Science and Technology 29, 207.
10. Sivashinsky, G. I., 1977, Combustion Science and Technology, 15, 137.
11. Matkowsky, B.J. and Olagunju, D.O. 1982, SIAM J. on Applied Mathematics, 42, 486.
12. Jackson, T. L. and Kapila, A. K., 1984, Combustion Science and Technology, 41, 191.
13. Kim, Y. D. and Matalon, M., 1987, submitted for publication.
14. Sivashinsky, G.I., Law, C.K. and Joulin, G. 1982, Combustion Science and Technology, 28, 155.
15. Tsuji, H. and Yamaoka, I., 1982, Nineteenth Symposium (Int.) on Combustion, The Combustion Institute, 1533.
16. Mendes-Lopes, J. M. C., 1983, Ph.D. thesis, Cambridge University.
17. Kim, Y. D. and Matalon, M., 1987, submitted for publication.
18. Ishizuka, S. and Law, C. K., 1982, Nineteenth Symposium (Int.) on Combustion, The Combustion Institute, 327.
19. Ishizuka, S., Miyasaka, K. and Law C. K., 1982, Combustion and Flame, 45, 293.
20. Ishizuka, S., 1987, private communication.

4. Discussion Sessions

EXTINCTION OF COUNTERFLOW DIFFUSION FLAMES WITH BRANCHING-TERMINATION CHAIN MECHANISMS: THEORY AND EXPERIMENT

C. K. Law
Department of Mechanical Engineering
University of California
Davis, California 95616

Abstract

The asymptotic structure and extinction of diffusion flames supported by a chain mechanism consisting of two two-body, thermoneutral, high-activation-energy branching reactions and a three-body, exothermic, zero-activation-energy termination reaction have been analyzed for the model problem of counterflow flames. A unique extinction criterion has been derived which contains a chain extinction limit and a global extinction limit; the latter is identified as Linan's extinction criterion for a one-step overall reaction. A companion experimental study shows that the density-weighted extinction strain rate increases only linearly with increasing pressure, thereby demonstrating weakened pressure dependence due to the influence of the three-body termination reaction.

1. INTRODUCTION

Analytical combustion modeling frequently invokes the approximation of a one-step overall reaction with a large activation energy. While such an approximation has proven to be extremely powerful and fruitful, it is however intrinsically incapable to describe the class of combustion phenomena characterized by multi-step chain branching and termination reactions, such as the pressure-temperature explosion limit of hydrogen/oxygen mixtures and the cool flame phemonena.

The fact that a branching-termination chain reaction scheme cannot be adequately approximated by a one-step overall large-activation-energy reaction can be appreciated by considering the activation, exothermicity, and pressure-sensitivity of the chain mechanism. First, we note that while the branching reactions are usually characterized by large activation energies, the termination reactions are temperature-insensitive and therefore their activation energies can be considered to be zero for all practical purposes. Thus activation of the termination reactions mainly depends on the availability of the radicals and is not directly controlled by the local temperature. Second, in a chain mechanism the dominant heat release steps are the termination reactions while the branching reactions are either endothermic or approximatly thermoneutral. Thus, from energetics point of view, it is inappropriate to associate large exothermicity with high activation, as is done in the one-step overall reaction approximation. Third, the intensity of a one-step overall reaction usually increases monotonically with pressure, being controlled by an overall reaction order which is frequently taken to assume a constant value close to two for reactions involving a fuel and an oxidizer. For a chain mechanism, however, with increasing pressure the fractional increase in the intensity of the three-body termination reaction is higher than that of the two-body branching reaction. The net effect is a progressively slower rate of increase of the overall reaction as pressure increases.

In order to demonstrate the importance of the branching-termination chain mechanism in flame modeling and to demonstrate their influence on the flame behavior, a model chain mechanism has been proposed and then applied to an analysis of the structure and extinction of diffusion flames [1]. Experiments have also been conducted [2] on the extinction of counterflow diffusion flames in response to pressure variations, with the objective of identifying the influence of the chain termination reaction on the overall reaction intensity. In the following the synopses of these works are presented and discussed from a unified viewpoint.

2. THEORY

The Kinetic Scheme

The model chain mechanism which captures the essential physics of chain branching and termination can be proposed as

$$F + R_1 \rightarrow 2R_2 \tag{1}$$

$$O + R_2 \rightarrow 2R_1 \tag{2}$$

$$R_1 + R_2 + M \rightarrow 2P + M \quad , \tag{3}$$

where F, O, P and M respectively designate fuel, oxidizer, product and a third body, while R_1 and R_2 are the radicals required to propagate the reaction scheme. Reactions (1) and (2) represent the irreversible, thermoneutral, high activation energy branching reactions which require radicals to proceed and which in turn generate more radicals. These two reactions are intimately coupled in that the product of the one becomes the reactant of the other. Reaction (3) represents the highly-exothermic, zero-activation energy, three-body termination reaction.

The reaction rates w_j' for reaction j, j = 1,2,3, can be respectively expressed as

$$w_1' = B_1'(p')^2 Y_F Y_{R1} \exp(-T_a'/T')$$ (4)

$$w_2' = B_2'(p')^2 Y_O Y_{R2} \exp(-T_a'/T')$$ (5)

$$w_3' = B_3'(p')^3 Y_{R1} Y_{R2} \quad ,$$ (6)

where B_j' is an appropriately-defined frequency factor, p' the pressure, T' the temperature, Y_i the mass fraction of species i, T_a' the activation temperature, and we have assumed that Reactions (1) and (2) have the same activation energy. Quantities with and without primes are dimensional and nondimensional respectively.

Conservation Equations

With conventional property and combustion assumptions, the nondimensional steady-state conservation equations for the concentrations of F, O, R_1 and R_1, and for energy T, can be respectively written as

$$L\{\tilde{Y}_F\} = - w_1$$ (7)

$$L\{\tilde{Y}_O\} = - w_2$$ (8)

$$L\{\tilde{Y}_{R1}\} = - w_1 + 2w_2 - w_3$$ (9)

$$L\{\tilde{Y}_{R2}\} = 2w_1 - w_2 - w_3$$ (10)

$$L\{T\} = w_3 \quad ,$$ (11)

where $L\{\cdot\}$ is the convective-diffusive differential operator. If we now specialize to counterflow combustion, with $z = z'/\sqrt{(D'/K')}$ being the nondimensional distance normal to the mixing layer measured from the stagnation plane, and if we also assume that the flow is incompressible and that the two streams have equal velocity, then it can be shown [3] that

$$L\{\cdot\} = - (\frac{d^2}{dz^2} - z\frac{d}{dz})\{\cdot\} \quad ,$$ (12)

with the boundary conditions,

$$z \rightarrow -\infty: \quad T = T_{-\infty} \quad , \quad \tilde{Y}_F = 1 \quad , \quad \tilde{Y}_O = \tilde{Y}_{R1} = \tilde{Y}_{R2} = 0 \tag{13}$$

$$z \rightarrow \infty: \quad T = T_{\infty} \quad , \quad \tilde{Y}_O = \alpha \quad , \quad \tilde{Y}_F = \tilde{Y}_{R1} = \tilde{Y}_{R2} = 0 \tag{14}$$

and the reaction rates

$$w_1 = \{Da_1(p^2/\rho K)\tilde{Y}_F\tilde{Y}_{R1}\}\exp\{-T_a/T\} \tag{15}$$

$$w_2 = \{Da_2(p^2/\rho K)\tilde{Y}_O\tilde{Y}_{R2}\}\exp\{-T_a/T\} \tag{16}$$

$$w_3 = \{Da_3(p^3/\rho K)\tilde{Y}_{R1}\tilde{Y}_{R2}\} \quad . \tag{17}$$

In the above $T = (c_p'/q_c')T'$, $\tilde{Y}_i$ is a stoichiometrically-weighted mass fraction, Da_j an appropriately-defined collisional Damköhler number, D' an average diffusivity, c_p' the specific heat, q_c' the specific heat of combustion, and K' the counterflow velocity gradient.

It is well known that by defining a new spatial coordinate

$$x = (1/2)\text{erfc}(z/\sqrt{2}) \quad , \tag{18}$$

the convection terms in the conservation equations can be eliminated while the flow domain is now bounded by $x = 0, 1$, which respectively correspond to $z = \infty$ and $-\infty$. Furthermore, using the coupling function formulation we find

$$\tilde{Y}_{R1} = T_{\infty} + 2\alpha - (2\alpha+\beta-1)x + \tilde{Y}_F - 2\tilde{Y}_O - T \tag{19}$$

$$\tilde{Y}_{R2} = T_{\infty} - \alpha + (\alpha-\beta+2)x + \tilde{Y}_O - 2\tilde{Y}_F - T \quad , \tag{20}$$

where $\beta = T_{\infty} - T_{-\infty}$. Thus only Eqs. (7), (8) and (11) need to be solved.

Asymptotic Solution and Flame Structure

The flame structure and response depend on the relative efficiency of the termination versus the branching reactions. Three types of flames have been identified, which are respectively characterized by fast, intermediate and slow recombinations [1]. We shall discuss only the flame structure of the fast recombination case which appears to be most relevant and interesting.

The flame structure is shown in Fig. 1. In the limit of $T_a \rightarrow \infty$, we have the Burke-Schumann flame-sheet solution with the flame temperature being that of the adiabatic flame temperature $T_{ad} = T_{\infty}+(1-\beta)x_f$, where $x_f = \alpha/(1+\alpha)$ is the flame location. For large but finite values of T_a such that $\varepsilon = T_{ad}^2/T_a \ll 1$, the flame sheet is broadened. In the outer zones bounded away from this reaction zone, the branching reactions are still effectively suppressed because of their temperature-sensitive Arrhenius kinetics, while the recombination reaction is suppressed due to the insufficient amount of radicals which are mostly consumed in the reaction zone. Within the reaction zone, the temperature peaks somewhere in the middle, while the concentrations of R_1 and R_2 respectively peak on the oxidizer and fuel side of it because of the need of these reactants to produce R_1 and R_2.

The maximum temperature in the reaction zone is $O(\varepsilon)$ smaller than T_{ad} while the concentrations of the reactants, F and O, are $O(\varepsilon)$. Furthermore, since during chemical reactions the concentrations of the radicals are usually much smaller than those of the reactants, we have imposed that the concentrations of the radicals R_1 and R_2 are $O(\varepsilon^2)$ and are therefore one order smaller than those of F and O. In the outer zones, the temperature and concentrations of F and O are $O(\varepsilon)$ smaller than their respective frozen values. The concentrations of R_1 and R_2, however, are $O(\varepsilon^2)$ because they have to be either smaller than or at most of the same order as their values in the reaction zone where they are produced.

Based on the above flame structure, the governing equations have been asymptotically analyzed. In the analysis the inner solution assumes the expansions

$$\{T^{in}, \tilde{Y}_F^{in}, \tilde{Y}_O^{in}\} = \{T_{ad}, 0, 0\}$$

$$+ (1/2)\varepsilon\delta^{-1/3}\{-[t(\xi)+\gamma\xi] , y_F(\xi) , y_O(\xi)\} \tag{21}$$

$$\{\tilde{Y}_{R1}^{in} , \tilde{Y}_{R2}^{in}\} = (1/2)\varepsilon^2\delta^{-1/3}\{\Delta_1^{-1}y_{R1}(\xi) , \Delta_2^{-1}y_{R2}(\xi)\} \tag{22}$$

$$\xi = (1+\alpha)\delta^{1/3}(x-x_f)/\varepsilon \quad , \tag{23}$$

where $\gamma = 1 - 2(1-\beta)/(1+\alpha)$,

$$\delta = \frac{\Delta_1\Delta_2}{\Delta_3} \tag{24}$$

is an effective Damköhler number of the branching-termination chain mechanism, and Δ_1, Δ_2, and Δ_3 are the Damköhler numbers for the individual reactions respectively given by

$$\Delta_1 = \frac{\varepsilon^4[Da_1(p^2/\rho_f K)f(x_f)\exp(-T_a/T_{ad})]}{2(1+\alpha)^2} = O(1) \tag{25}$$

$$\Delta_2 = \frac{\varepsilon^4[Da_2(p^2/\rho_f K)f(x_f)\exp(-T_a/T_{ad})]}{2(1+\alpha)^2} = O(1) \tag{26}$$

$$\Delta_3 = \frac{\epsilon^5 [Da_3 (p^3/\rho_f K) f(x_f)]}{2(1+\alpha)^2} = O(1) \qquad , \tag{27}$$

with $f(x) = 2\pi \exp(z^2)$.

The asymptotic solution thus yields the flame structure described by the following expressions:

$$y_F = t + \xi \tag{28}$$

$$y_O = t - \xi \tag{29}$$

$$y_{R1} = (t-\xi) \exp\{-(1/2)\delta^{-1/3}(t+\gamma\xi)\} \tag{30}$$

$$y_{R2} = (t+\xi) \exp\{-(1/2)\delta^{-1/3}(t+\gamma\xi)\} \qquad , \tag{31}$$

where t is given by

$$\frac{d^2 t}{d\xi^2} = (t^2 - \xi^2) \exp\{\delta^{-1/3}(t+\gamma\xi)\} \tag{32}$$

$$(\frac{dt}{d\xi})_{-\infty} = -1 \quad , \quad (\frac{dt}{d\xi})_{\infty} = 1 \quad . \tag{33}$$

Equation (32) subject to the boundary conditions of Eq. (33) is in exactly the same form as that describing the near-equilibrium diffusion flame structure of Linan [3] in which a one-step overall reaction is assumed. Thus Linan's solution for $t(\xi)$ can be directly used. In particular, it is shown [3] that for a given γ there exists a critical δ, called δ_E, such that two solutions exist for $\delta > \delta_E$ and there is no solution for $\delta < \delta_E$. Thus δ_E can be identified as an extinction δ in that extinction is expected to take place for systems satisfying

$$\delta < \delta_E \quad . \tag{34}$$

Furthermore, Linan has shown that δ_E can be explicitly correlated to γ according to

$$\delta_E(\gamma) = e[(1-\gamma) - (1-\gamma)^2 + 0.26(1-\gamma)^3 + 0.055(1-\gamma)^4] \quad . \tag{35}$$

When $\gamma < 0$, $|\gamma|$ is to be used.

The present model contains more information than that of Linan [3] even though in reduced form these results are identical. Specifically, we shall show in the following that our model can be interpreted to consist of a chain extinction limit and a global extinction limit.

Extinction Limits

In the chain extinction limit we compare the relative efficiencies of the branching versus the termination reactions. For example, if we hold the Damköhler numbers Δ_1 and Δ_2 of the branching reactions fixed, then Eq. (24) and the general extinction criterion (34) clearly show that there exists a (chain) extinction Damköhler number $\Delta_{3,E}^{ch}$ for the termination reaction, given by

$$\Delta_{3,E}^{ch} = \frac{(\Delta_1 \Delta_2)}{\delta_E} \quad , \tag{36}$$

such that <u>chain extinction</u> is expected to occur if

$$\Delta_3 \gtrsim \Delta_{3,E}^{ch} \quad . \tag{37}$$

In this limit the termination reaction overwhelms the branching reactions by removing the radicals needed for the branching reactions.

The above result can be alternatively interpreted by varying one or both of the branching Damköhler numbers Δ_1 and Δ_2 while holding the rest of the Damköhler numbers fixed. For example, if we hold Δ_3 fixed and vary both Δ_1 and Δ_2, then we get a combined extinction Damköhler number for the branching reactions,

$$(\Delta_1 \Delta_2)_E^{ch} = \Delta_3 \delta_E \tag{38}$$

such that chain extinction is expected to occur if

$$(\Delta_1 \Delta_2) \lesssim (\Delta_1 \Delta_2)_E^{ch} \quad . \tag{39}$$

In the global extinction limit the branching and termination reactions occur in parallel such that

$$(\frac{\Delta_3}{\Delta_1}) = m = O(1) \tag{40}$$

$$(\frac{\Delta_3}{\Delta_2}) = n = O(1) \quad . \tag{41}$$

Applying Eqs. (40) and (41) into Eq. (24), and using the general extinction criterion (34), it can be stated that there exists a (global) extinction Damköhler number $\Delta_{3,E}^{gl}$ for the termination reaction, given by

$$\Delta^{gl}_{3,E} = (mn)\delta_E \qquad (42)$$

such that <u>global extinction</u> is expected to occur if

$$\bar{\Delta}_3 \lessapprox \Delta^{gl}_{3,E} \qquad . \qquad (43)$$

This result is identical to that of Linan, hence the term global extinction.

It is important to note the qualitatively opposite influence of Δ_3 on the chain versus the global extinction criteria. That is, for chain extinction, <u>increasing</u> Δ_3 while holding Δ_1 and Δ_2 fixed facilitates extinction. On the other hand, for global extinction <u>decreasing</u> Δ_3 while holding (Δ_3/Δ_1) and (Δ_2/Δ_1) fixed facilitates extinction.

3. EXPERIMENT

A direct investigation of the two extinction limits discussed in the previous section cannot be readily conducted because of the uncertainty in identifying the representative branching and termination reactions. However, the extinction criterion as given by the extinction Damköhler number does provide explicit functional relations between the system parameters amenable for experimental exploration.

Experiments were conducted in a counterflow burner in which two axisymmetric, nozzle-generated gaseous streams impinge onto each other. For given concentrations of the fuel and oxidizer streams, the extinction strain rate of the diffusion flame is determined by continuously increasing the stream velocities until extinction occurs. The strain rate is defined as half of the constant axial velocity gradient upstream of the thermal diffusion zone of the flame, and is determined by using laser Doppler velocimetry. The burner is housed in a chamber with continuous flow and a maximum pressure of about five atmospheres.

In order to demonstrate the importance of the branching-termination chain mechanism in flame extinction, we need to enhance the intensity of the termination reaction relative to that of the branching reaction. This can be achieved by reducing the flame temperature, which weakens the temperature-sensitive branching reactions, and increasing the system pressure, which facilitates the three-body termination reaction. In the present experiments the flame temperatures were reduced by diluting the fuel stream by nitrogen.

Figure 2 plots the extinction strain rate K'_{ex} as a function of pressure for a diffusion flame of CH_4/N_2 versus air, with 18.7 and 25% of CH_4 in the CH_4/N_2 stream. It is seen that with increasing pressure K'_{ex} decreases monotonically for the weaker 18.7%-CH_4 flame, but increases and then decreases for the 25%-CH_4 flame.

An inspection of the Damköhler number definition, given by Eqs. (25) to (27), shows that the relevant strain rate should be $\rho_f' K'$ instead of K'. Furthermore, since $\rho_f' \sim \rho_\infty'$ for a fixed mixture concentration, the dependence of straining on pressure to achieve extinction can be identified by plotting $\rho_\infty' K_{ex}'$ versus p', as shown in Fig. 3. It is seen that $\rho_\infty' K_{ex}'$ increases monotonically with p'. Furthermore, this dependence can be approximately correlated by

$$\rho_\infty' K_{ex}' \sim p' \quad . \tag{44}$$

The value of the proportionality constant decreases with decreasing flame temperature and thereby burning intensity.

The linear functional form of (44) agrees with our theoretical result. That is, by substituting the extinction values of Δ_1, Δ_2, and Δ_3 of Eqs. (25) to (27) into Eq. (24), we obtain identically (44). Furthermore, the proportionality constant should depend mostly on $\exp(-2T_a/T_{ad})$, which decreases with decreasing T_{ad} and thereby increasing N_2 dilution. This again agrees with the experimental result.

It is necessary to clarify that the agreement in the linear pressure dependence is actually somewhat fortuitous because we have assumed that (1) to (3) are elementary reactions when writing the reaction rates w_j' in Eqs. (4) to (6). Since (1) to (3) are actually model reactions, we could have used a general, overall reaction order of n_j for each reaction rate w_j'. This will lead to a theoretical dependence of

$$\rho_\infty' K_{ex}' \sim (p')^{(n_1 + n_2 - n_3)} \tag{45}$$

which degenerates to (44) only for $n_1 = n_2 = 2$ and $n_3 = 3$.

Since a near-quadratic pressure dependence would have resulted in the density-weighted extinction strain rate had we used a one-step overall reaction, the present experimental result of linear dependence does demonstrate a weakened pressure influence on the burning intensity because of the presence of the termination reaction.

4. CONCLUDING REMARKS

In the present investigation we have demonstrated the importance of allowing for multi-step branching-termination chain mechanisms for satisfactory description of the structure and extinction of diffusion flames under certain situations. Recent experimental studies [2] have further shown that chain mechanisms can result in negative values of overall reaction orders and are also responsible for the phenomena of flammability limits. It is clear that more work is needed on the role of chain mechanisms in theoretical combustion studies.

Acknowledgement

It is a pleasure to acknowledge the support of the work summarized herein by the Air Force Office of Scientific Research and by the Division of Basic Energy Sciences, Department of Energy.

References

1. Birkan, M. A. and Law, C. K., "Asymptotic Structure and Extinction of Diffusion Flames with Chain Mechanisms," to appear in Combustion and Flame, 1988.
2. Law, C. K. and Egolfopoulos, F. N., "Chain Mechanisms in the Propagation/ Extinction of Flames and the Determination of Flammability Limits," to be published.
3. Linan, A., "The Asymptotic Structure of Counterflow Diffusion Flames for Large Activation Energies," Acta Astronautica 1, 1007-1039 (1974).

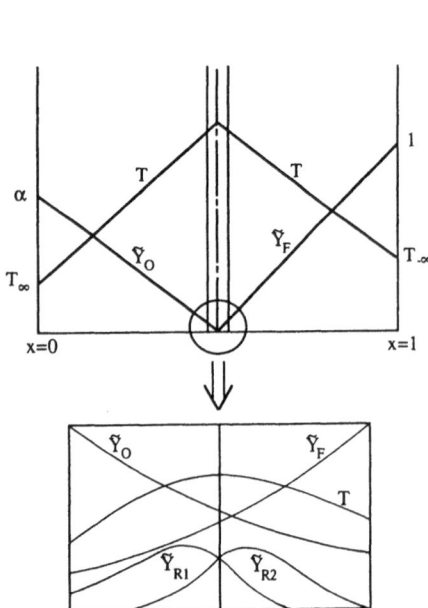

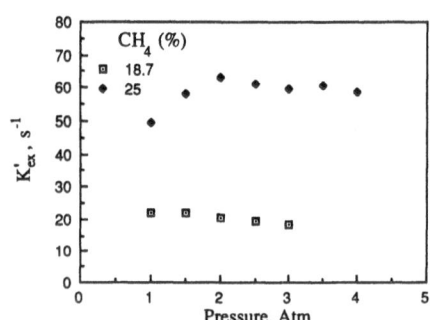

Figure 2   Extinction strain rates as function of pressure for $CH_4/$ N_2 and air diffusion flames.

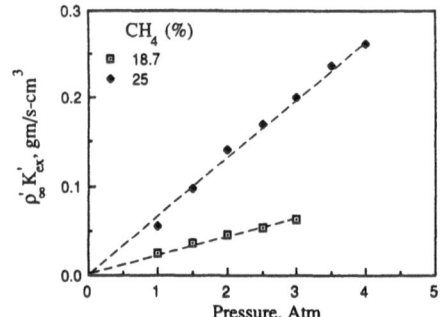

Figure 1   Temperature and concentration profiles in the flame-sheet limit and within the reaction zone.

Figure 3   Density-weighted extinction strain rates as function of pressure for CH_4/N_2 and air diffusion flames.

SOME WORKSHOP TOPICS

J. Buckmaster
University of Illinois
Urbana, IL 61801

Abstract

In addition to their substantial formal presentations, workshop participants were given opportunities for additional talks, and for brief 'workshop' presentations, comments, etc.. Here is a brief description of some of those topics discussed in which I -and others- were involved.

Pressure Transients and the Genesis of Transverse Shocks in Unstable Detonations

A paper with this title will be published elsewhere, [1]. Its abstract is:

'We examine unstable overdriven detonations with one-step Arrhenius' kinetics in the limit of infinite activation energy. For the one-dimensional problem we show how a certain class of initial disturbances lead to thermal runaway for the entire detonation structure, with the shock temperature and pressure increasing several-fold in an extremely small time interval. This behavior resembles the sharp transients reported by others in numerical computations. For the two-dimensional problem, runaway occurs at discrete points distributed along the plane of the shock with spacing determined by the initial disturbance. This description is characterized by very large transverse pressure gradients which, in turn, will give rise to the transverse shock waves that are a familiar feature of unstable multidimensional detonations.'

Hydrogen Flame Bubbles - a possible experiment

Elsewhere in these proceedings is a paper 'Convection effects and the stability of hydrogen flame bubbles'. There it is pointed out that an important ingredient of the flow field associated with these flames is the wake, in which, because of buoyancy forces, there is a persistent velocity defect. The magnitude of this defect depends on the Froude number and the heat released at the flame. Experimental verification of the wake would be worthwhile, and perhaps this could be achieved by measuring the velocity along the center-line of a rising 'bubble'. The only non-graphic experimental information on these flames is the rise speed, so that any other quantitative knowledge would be valuable.

Wind Induced by Diffusion Flame Extinction - a possible experiment

Reference [2] describes what happens to a diffusion flame when the Damkohler number is decreased, so that extinction occurs. The analysis is valid when the reaction is modeled by one-step, irreversible, large-activation-energy kinetics. If the non-dimensional activation energy is θ, the thickness of the flame sheet, or reaction zone, is $O(\theta^{-1})$, and the spatial temperature variations within it are $O(\theta^{-1})$ so that the diffusion term, $\partial^2 T/\partial n^2$ is $O(\theta)$. Thus a natural time scale for unsteady changes in the zone is one for which $\partial T/\partial t$ is $O(\theta)$, i.e., $\partial/\partial t = O(\theta^2)$. The initial stage of the extinction process occurs on this very short time scale, and is responsible for the rapid generation of an $O(1)$ velocity field in which fluid is drawn into the reaction zone from outside. Only very small changes in the temperature field occur during this process. It would be of interest to measure these transients experimentally.

An interesting question is whether the fast time scale plays a role in unsteady premixed flames. If the stationary flame sheet structure is unstable, e.g. [3], the answer is yes. But in the extinction of stable structures there is a fundamental difference between premixed and diffusion flames. Damkohler number extinction of diffusion flames is associated with non-existence of a stationary solution of the structure equation. For premixed flames, the structure equation always has a stationary solution, and extinction (by heat losses, for example) is associated with the nonexistence of a stationary solution of the combustion field beyond the flame sheet. In this situation, the fast time will play no role.

Detonation Instabilities - a possible experiment

Existing experimental observations of unstable detonation waves reveal an apparent dichotomy. In the case of detonations traveling down tubes, there are transverse waves which generate the familiar diamond pattern on witness plates, [4]. For detonations as bow waves on hypersonic blunt bodies, one-dimensional pulsations are observed, [5]. It isn't clear whether these are separate modes of instability, or a single mode for which some characteristics are suppressed in the blunt body configuration. The blunt body experiments appear only to have been carried out using rounded bodies (spheres). Perhaps if flat faced bodies were used of varying thickness, a transition between one-dimensional pulsations and the appearance of transverse waves could be observed.

There is an important related theoretical question. Most of our knowledge of linear detonation stability arose from the work of Erpenbeck (described in [6]). For the most part this is restricted to a description of the stability boundaries. Little is known of the nature of the spectrum, and yet this must play an important role in the expected manifestation of any instability. In view of the much more powerful computational tools available today, compared to what was available when Erpenbeck did his work, a detailed re-examination of this problem seems very desirable.

References

[1]   J. Buckmaster, Combustion Science and Technology, to appear.

[2]   J. Buckmaster, D. S. Stewart and A. Ignatiadis, Combustion Science and Technology, 46, 145-165 (1986).

[3]   D. S. Stewart, Combustion and Flame, 64, 157-165 (1986).

[4]   R. A. Strehlow, Combustion Fundamentals, McGraw-Hill, New York, 1984, p. 311.

[5]   R. L. Alpert and T. Y. Toong, Astronautica Acta, 17, 539-560 (1972).

[6]   W. Fickett and W. C. Davis, Detonation, University of California Press, Berkeley, CA, 1979.

ASYMPTOTIC APPROACH TO ANALYSIS OF PROPELLANT COMBUSTION

F.A. Williams
Department of Mechanical & Aerospace Engineering
Princeton University
Princeton, NJ 08544

INTRODUCTION

In recent years there have been various applications of asymptotic methods to problems in propellant combustion. Most of these have involved activation-energy asymptotics. Usually the analyses have been generic in character, with the objective of uncovering types of phenomena that may arise. An example is a recent analysis of influences of two-phase flow, such as bubbling of a reacting liquid, on deflagration velocities of propellants[1]. A few analyses have addressed burning of real propellants and have attempted to predict burning velocities for comparison with experiment. A representative example is the recent analysis of the deflagration of nitramines and nitramine propellants, which achieved good agreement with measured burning velocities and their pressure and temperature sensitivities[2]. This latter analysis is reviewed to illustrate the character of these studies. Since asymptotic methods recently have achieved success in describing the deflagration of gaseous fuels, with account taken of detailed chemistry, the question is raised as to whether the time is ripe for addressing detailed chemistry in propellant combustion by these approaches.

NITRAMINE DEFLAGRATION

Nitramines are crystals that melt, decompose exothermically in the liquid phase, gasify and decompose further in the gas. Bubbling is observed in the liquid, and although the adiabatic flame temperature is T_a = 3200 K, the gas-phase combustion is observed to occur in two stages, with only the first stage, which is completed at a temperature of about T_f = 1330 K, affecting the deflagration velocity. The recent analysis of nitramine deflagration[2] employed activation-energy asymptotics for the first-stage, gas-phase reaction and also for the liquid-phase reaction, hypothesizing evaporative equilibrium at the liquid-gas interface. The second stage of the gas-phase reaction was ignored completely because of its absence of any influence on the burning velocity. The gas-phase reaction that was included was approximated as a one-step, Arrhenius process, with an overall activation energy E = 32.5 kcal/mol and a pressure exponent n = 1.6 for the prefactor, the latter adjusted empirically to provide best fits to available data. On the other hand, the overall activation energy for the condensed-phase reaction (which was assigned no pressure dependence), was selected to be the energy required to break the N–N bond, on the basis of analysis of available decomposition-rate data, evaluation of their significance, and estimation of the enhanced importance of the initiation step at the hotter, faster conditions encountered in deflagration.

The theory was successful not only with respect to the deflagration rate, but also in identifying the low-pressure deflagration limit with a condition of interface adiabaticity, seen by the condensed phase at its surface. With this adiabaticity, the condensed-phase reaction can propagate ahead, independent of the gas, but this mechanism is known to be subject to strong pulsating instability that could lead to extinction. In fact, pulsations are observed in nitramines near the deflagration limit. Thus, an unplanned success, not built into the model, lent enhanced credence to the theory. It may be remarked that in the original work, some adjustment in overall rate levels from average experimental results were made to improve deflagration-velocity agreement[2], but with account taken of bubbling[1] (which had been excluded by treating the liquid-gas interface as planar[2]), these adjustments no longer were needed.

CRITIQUE OF THEORY

The view may be taken that the theory is a tremendous success because its results agree with experiment even in ways not originally designed. Alternatively, the view may be taken that the theory is a terrible failure because if offers no insight whatever about what the true gas-phase chemistry may be; it simply makes totally empirical choices of n and T_f to achieve agreement. A balanced evaluation undoubtedly falls somewhere between these extremes. Of greater interest is the question of whether the theory can be improved by paying more attention to detailed chemical kinetics. Since we now have ways to employ asymptotic methods for gaseous deflagrations with detailed chemistry, shouldn't we try the same kind of thing for the condensed-phase and gas-phase kinetic processes in propellant combustion? Mightn't this lead to a more satisfactory theory of nitramine deflagration?

APPRAISAL OF CHANCES OF SUCCESSFUL INCLUSION OF DETAILED CHEMICAL KINETICS

The answers to the preceding questions remain unclear. In most propellant combustion processes, the chemical kinetics are potentially much more complex than those of gaseous deflagrations, and the rates of elementary steps are known much less accurately. For example, it is uncertain whether sufficient chemical-kinetic information is available even to predict the existence of two-stage gas-phase combustion in nitramine deflagration. Further investigations of gas-phase chemical kinetics may help to clarify these questions, but this is difficult because of the complexity of the key kinetic steps. These early-stage gas-phase kinetics, and especially condensed-phase kinetics, typically involve larger and more complicated molecules whose reaction channels are poorly understood. It would be difficult to have confidence in applications of asymptotic methods based on selected elementary rate parameters that are so uncertain. thus, prospects for success in applications of asymptotics with detailed chemistry to propellant deflagration do not seem high. The chemistry is just too complicated; the most useful level of attack seems still to involve physically motivated empiricism. It would be of interest to pursue asymptotic methods with detailed chemistry for propellant deflagration, but mainly for the systems of this type they exhibit the simplest possible chemical kinetics. Tests with simpler systems could give better insight into prospects for success with more complicated systems.

REFERENCES

1.   S.B. Margolis, F.A. Williams and R.C. Armstrong, "Influence of Two-Phase Flow on the Deflagration of Homogeneous Solids," Combustion and Flame 67, 249-258 (1987).

2.   T. Mitani and F.A. Williams, "A Model for the Deflagration of Nitramines," Twentieth Symposium (International) on Combustion, The Combustion Institute, Pittsburgh, to appear (1987).

ONSET OF INSTABILITY IN A SPHERICALLY GROWING FLAME

M. Matalon
Engineering Sciences and Applied Mathematics
The Technological Institute
Northwestern University
Evanston, IL 60208

In stability theory, one begins with a simple laminar flow, assumes the presence of disturbances at time $t = 0$, and follows their development with time. If as time goes on, the basic flow without disturbance tends to reemerge, then it is said to have "stability" (strictly speaking with respect to the particular disturbance assumed). If the basic state becomes more and more deformed by the growth of the disturbance, it is said to have "instability". Thus, stability or instability is determined by the asymptotic behavior of the disturbances as $t \to \infty$. Clearly, this is meaningful if the basic flow is steady. For an unsteady basic flow the behavior at a _finite_ time is also of primary interest, so that the tendency toward stability or instability at any moment may be relevant [1]. In order to determine that tendency it seems natural to compare the disturbance growth rate with the rate of change of the basic state.

For definiteness, consider a spherically symmetric flame originating from an ignition point source and propagating outwardly in a combustible mixture. According to the arguments stated above, if at a given instant the disturbance increases but the flame (the basic state) grows even more rapidly, then the disturbance would appear to be decaying and the flame is momentarily stable. Conversely, if the disturbance grows at the faster rate, the flame is momentarily unstable. The terms instability/stability are thus used in a special sense reflecting only the tendency of the spherical flame front to become more and more or less and less distorted. Now, if under certain conditions the flame is "momentarily unstable" for all $t > t_0$, an instability will result and will be observed at a time $t_c \geq t_0$. How to determine t_c is not clear.

There have been several observations of spherical flames exhibiting cellular instability. In one of the most detailed study [4], a lean propane-air mixture was ignited at the center of a 13 cm radius spherical, constant volume vessel. A smooth spherical flame was first observed, but when its radius was about 7 - 10 cm, it took on a cellular appearance. Photographs showing the development of the flame were taken at time interval of about

12 ms. Based on these photographs the various parameters at the onset of the cellular instability were determined. In particular, the estimated critical Reynolds number was in the range of 4000 - 9000 and the spherical harmonic at the onset of the instability in the range of 60 - 100.

Theoretical studies [2],[3] on the other hand have been limited to spherically growing flames in an infinite space. Following the evolution of small disturbances it is found that the amplitude of a disturbance, measured relative to the growing flame size, behaves as in Figure 1. This result [3] corresponds to a mixture for which the Lewis number Le, based on the molecular diffusivity of the deficient reactant, is sufficiently large or, Le > Le* with Le* slightly less than unity, which holds for example in a lean propane-air mixture as in the experiment reported earlier [4]. Disturbances are introduced at time t - 0 when the flame size is R - 1; prior to that time the flame size was comparable to the diffusion length and the flame was absolutely stable since diffusional and thermal effects have stabilizing influence for Le > 1. As the flame expands, the relative amplitude first decreases with time reaching a minimum at $R = R_0$ (or $t = t_0$) and then begins

to grow. For parameter values corresponding to a lean propane-air mixture the critical radius R_0 was found to correspond to a Reynolds number of

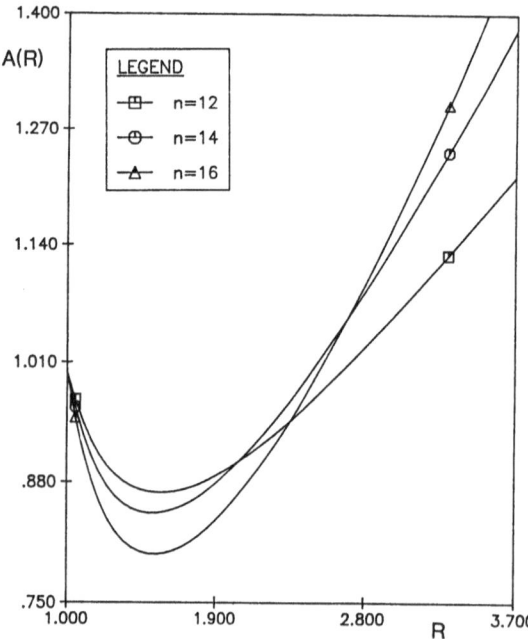

Fig. 1 The evolution of the amplitude A of a disturbance, measured relative to the growing flame size. Here R is the unperturbed flame front position (R ~ t) and n is the spherical harmonic. This result is based on the calculation of Ref. [3].

about 1000. Disturbances with spherical harmonic $n - n_0$ where $n_0 \approx 14$ appear to reach their minimum first and therefore seem to be the most destabilizing at $t = t_0$. This is when instability is assumed to occur and the flame surface is immediately covered with a large number of cells corresponding to n_0.

A direct comparison of the Reynolds number calculated from the theory at time t_0, and that estimated from the experimental data at time when the cells were first observed, show that the latter is larger by a factor of 5 to 10. Although in the experiment the flame was expanding in a confined vessel and was therefore subjected to a pressure buildup, one of the cases reported has the flame becoming cellular before the pressure has increased by 20%. Thus, the different hydrodynamic conditions cannot provide a complete explanation for the discrepancy between the theoretical and experimental estimates. The question appears to be how to relate information regarding the development of the amplitude of a disturbance with observations of the type reported in [4]. In other words, what is the relation between t_0 and the time when the instability is first observed in an experiment? One could argue that, to be observed, the amplitude of the unstable mode must grow in size to the magnitude it had when first introduced, or maybe several times this magnitude. Indeed the significantly larger Reynolds number estimated from the experimental data suggests that the transition to the cellular structure may have occurred at a time $t_c > t_0$. And the question remains how to determine t_c. Therefore, experiments which examine more carefully the onset of the instability and the evolution of the cellular structure would be very useful in addressing this question.

As a final comment, it should be emphasized that the questions raised in this note are not peculiar to flame propagation but relevant as well to hydrodynamic stabilities of monotonic unsteady flows. This is in contrast to the stability of time-periodic flows [5] where the growth/decay over each modulation cycle can be used in order to determine stability or instability.

References

1. Shen, S.F., 1961, Journal of Aerospace Sciences, **28**; 397.
2. Istratov, A.G., and Librovich, V.B., 1969, Astronautica Acta, **14**, 453.
3. Bechtold, J.K., and Matalon, M., 1987 Combustion and Flame, **67**, 77.
4. Groff E.G., 1982, Combustion and Flame, **48**, 51.
5. Davis S. H., 1976, Annual Review of Fluid Mechanics, **8**, 57.

Mg/TF Propellant Combustion

Naminosuke Kubota

Third Research Center, Technical Research and Development Institute
Japan Defense Agency

Introduction

There have been very limited studies on the combustion of the propellants consisted of metal particles and oxidizers. The burning rate characteristics of this class of propellants are dependent on various physical and chemical parameters. Since the physical structure of the propellants is heterogeneous, the combustion wave structure appears to be highly heterogeneous. The diffusional mixing of the gaseous and condensed fuel/oxidizer fragments occurs on and above the burning surface of the propellants.

In this study, the combustion process of Mg(magnesium)/TF(tetrafluoroethylene) propellants was examined in order to gain informations which control the burning rate characteristics.[1] TF is composed of $-C_2F_4-$ molecular structure which contains 0.75 weight fraction of fluorine. The heat produced by the oxidation of Mg with fluorine is 16.8 MJ/kg.

Burning Rate Characteristics

The Mg/TF propellants were made as pressed pellets which consisted of various sizes and concentrations of Mg particles. Figure 1 shows the effect of the weight fraction of Mg(ξ) on burning rate (r) and the adiabatic flame temperature (T_f). The burning rate increases monotonically with increasing ξ, whereas T_f increases with increasing ξ in the region $\xi < 0.33$ and decreases with increasing ξ in the region $\xi > 0.33$. It must be noted that the burning rate increases drastically even though T_f decreases in the region $\xi > 0.33$. The effects of the Mg particle size (θ) and the total surface area of the Mg particles mixed within the unit mass of the propellant (Σ) were also measured. The burning rate increases with decreasing θ at a constant ξ. Furthermore, the burning rate increases linearly in a logarithmic burning rate versus Σ plot as shown in Fig. 2.

Combustion Wave Structure

The temperature distribution in the combustion wave was measured with micro-thermocouples. The temperature increases from the initial propellant temperature (T_0) to the burning surface temperature (T_s), and continues to increase rapidly in the gas phase above the burning surface. As shown in Fig. 3, T_s decreases with increasing ξ, whereas the burning rate increases with increasing ξ.

If one assumes a one-dimentional combustion wave along the burning direction, the overall heat balance at the burning surface is represented by

$$r = \alpha_s \phi / \psi \qquad (1)$$

where
$$\phi = (dT/dx)_{s,g} \qquad (2)$$
$$\psi = T_s - T_0 - Q_s/c_p \qquad (3)$$
$$\alpha_s = \lambda_g/c_p \rho_p \qquad (4)$$

T is temperature, x is distance, Q_s is the heat of reaction at the burning surface, λ is thermal conductivity, ρ is density, c is specific heat, and the subscripts g is gas phase, p is propellant, and s,g is the gas phase at the burning surface.

Figure 4 shows the calculated results of Q_s as a function of ξ. The negative value of Q_s is caused by the melt of the Mg particles (heat of fusion is -379 kJ/kg) and the decomposition of the TF (heat of decomposition is -6580 kJ/kg) at the propellant burning surface. The heat flux transferred back from the gas phase to the burning surface (Λ) is given by

$$\Lambda = \lambda_g \phi \qquad (5)$$

As also shown in Fig. 4, Λ increases monotonically as ξ increases. These results indicate that the reaction rate in the gas phase increases with increasing ξ. In the computations of Q_s and Λ, the data of burning rate shown in Fig. 1 are used. The physical parameter values used are : $\rho_p = 1.8 \times 10^3$ kg/m^3 and $c_p = 1.05$ kJ/kgK.

Based on the results obtained in this study, the following reaction schema of Mg/TF propellants is represented: The melted Mg particles on the propellant burning surface are ejected into the gas phase. In the gas phase, the oxidation of the Mg particles by the fluorine produced by the decomposition of the TF occurs from the surface of each Mg particle toward inside of the particle. The reaction completes at the far-downstream of the propellant burning surface. Since the available fuel component at the burning surface is a thin surface-layer surrounding each Mg particle, the fuel/oxidizer ratio approaches toward stoichiometric ratio as Σ increases. Thus, the reaction rate in the gas phase and the heat feedback process to the propellant are attributed to the effective stoichiometry of the reaction between the surface-layers of the Mg particles and the fluorine.

The burning rate behavior of Mg/TF propellants demonstrates a significant contrast when compared with conventional solid rocket propellants. The burning rate of solid rocket propellants increases with increasing T_f. In conclusions, the oxidation process of the melted Mg particles with fluorine in the steep temperature gradient just above the burning surface should be understood in order to describe the observed anomalous burning rate behavior of the Mg/TF propellants. The mathematical modeling of ϕ is the major task of the combustion study of this class of propellants.

Acknowledgment

The author wishes to thank Dr. T. Niioka who presented this subject at Joint USA/Japan Seminar on Mathematical Modeling in Combustion Science.

References

1. Kubota, N. and Serizawa, C., "Combustion of Magnesium/Polytetrafluoroethylene," J. of Propulsion and Power, Vol. 3, No. 4, 1987, pp. 303-307.

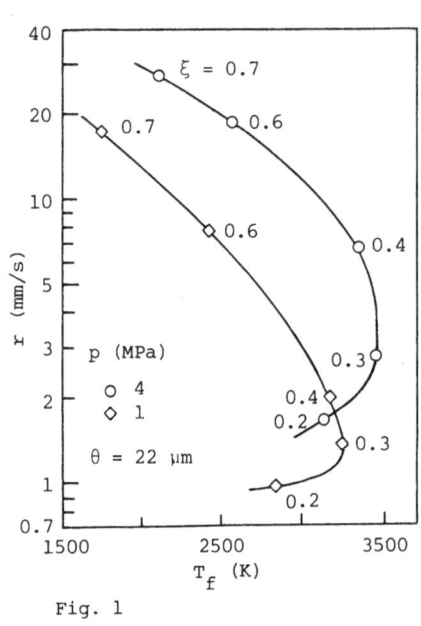

Fig. 1

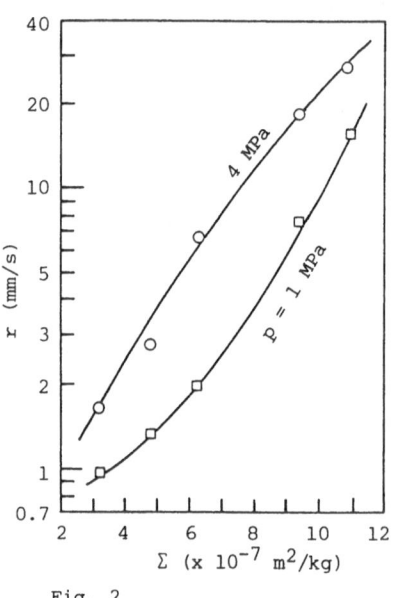

Fig. 2

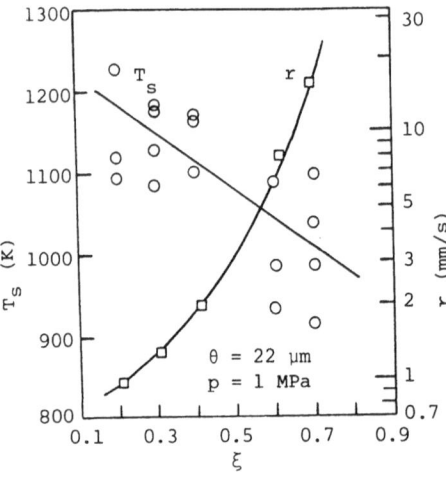

Fig. 3

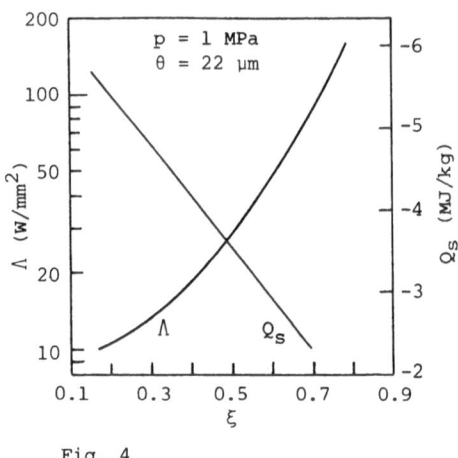

Fig. 4

Lecture Notes in Mathematics

Vol. 1174: Categories in Continuum Physics, Buffalo 1982. Seminar. Edited by F.W. Lawvere and S.H. Schanuel. V, 126 pages. 1986.

Vol. 1184: W. Arendt, A. Grabosch, G. Greiner, U. Groh, H.P. Lotz, U. Moustakas, R. Nagel, F. Neubrander, U. Schlotterbeck, One-parameter Semigroups of Positive Operators. Edited by R. Nagel. X, 460 pages. 1986.

Vol. 1186: Lyapunov Exponents. Proceedings, 1984. Edited by L. Arnold and V. Wihstutz. VI, 374 pages. 1986.

Vol. 1187: Y. Diers, Categories of Boolean Sheaves of Simple Algebras. VI, 168 pages. 1986.

Vol. 1190: Optimization and Related Fields. Proceedings, 1984. Edited by R. Conti, E. De Giorgi and F. Giannessi. VIII, 419 pages. 1986.

Vol. 1191: A.R. Its, V.Yu. Novokshenov, The Isomonodromic Deformation Method in the Theory of Painlevé Equations. IV, 313 pages. 1986.

Vol. 1194: Complex Analysis and Algebraic Geometry. Proceedings, 1985. Edited by H. Grauert. VI, 235 pages. 1986.

Vol. 1203: Stochastic Processes and Their Applications. Proceedings, 1985. Edited by K. Itô and T. Hida. VI, 222 pages. 1986.

Vol. 1209: Differential Geometry, Peñíscola 1985. Proceedings. Edited by A.M. Naveira, A. Ferrández and F. Mascaró. VIII, 306 pages. 1986.

Vol. 1214: Global Analysis – Studies and Applications II. Edited by Yu.G. Borisovich and Yu.E. Gliklikh. V, 275 pages. 1986.

Vol. 1218: Schrödinger Operators, Aarhus 1985. Seminar. Edited by E. Balslev. V, 222 pages. 1986.

Vol. 1227: H. Helson, The Spectral Theorem. VI, 104 pages. 1986.

Vol. 1229: O. Bratteli, Derivations, Dissipations and Group Actions on C*-algebras. IV, 277 pages. 1986.

Vol. 1236: Stochastic Partial Differential Equations and Applications. Proceedings, 1985. Edited by G. Da Prato and L. Tubaro. V, 257 pages. 1987.

Vol. 1237: Rational Approximation and its Applications in Mathematics and Physics. Proceedings, 1985. Edited by J. Gilewicz, M. Pindor and W. Siemaszko. XII, 350 pages. 1987.

Vol. 1250: Stochastic Processes – Mathematics and Physics II. Proceedings 1985. Edited by S. Albeverio, Ph. Blanchard and L. Streit. VI, 359 pages. 1987.

Vol. 1251: Differential Geometric Methods in Mathematical Physics. Proceedings, 1985. Edited by P.L. García and A. Pérez-Rendón. VII, 300 pages. 1987.

Vol. 1255: Differential Geometry and Differential Equations. Proceedings, 1985. Edited by C. Gu, M. Berger and R.L. Bryant. XII, 243 pages. 1987.

Vol. 1256: Pseudo-Differential Operators. Proceedings, 1986. Edited by H.O. Cordes, B. Gramsch and H. Widom. X, 479 pages. 1987.

Vol. 1258: J. Weidmann, Spectral Theory of Ordinary Differential Operators. VI, 303 pages. 1987.

Vol. 1260: N.H. Pavel, Nonlinear Evolution Operators and Semi-groups. VI, 285 pages. 1987.

Vol. 1263: V.L. Hansen (Ed.), Differential Geometry. Proceedings, 1985. XI, 288 pages. 1987.

Vol. 1267: J. Lindenstrauss, V.D. Milman (Eds), Geometrical Aspects of Functional Analysis. Seminar. VII, 212 pages. 1987.

Vol. 1269: M. Shiota, Nash Manifolds. VI, 223 pages. 1987.

Vol. 1270: C. Carasso, P.-A. Raviart, D. Serre (Eds), Nonlinear Hyperbolic Problems. Proceedings, 1986. XV, 341 pages. 1987.

Vol. 1272: M.S. Livšic, L.L. Waksman, Commuting Nonselfadjoint Operators in Hilbert Space. III, 115 pages. 1987.

Vol. 1273: G.-M. Greuel, G. Trautmann (Eds), Singularities, Representation of Algebras, and Vector Bundles. Proceedings, 1985. XIV, 383 pages. 1987.

Lecture Notes in Physics

Vol. 278: The Physics of Phase Space. Proceedings, 1986. Edited by Y.S. Kim and W.W. Zachary. IX, 449 pages. 1987.

Vol. 279: Symmetries and Semiclassical Features of Nuclear Dynamics. Proceedings, 1986. Edited by A.A. Raduta. VI, 465 pages. 1987.

Vol. 280: Field Theory, Quantum Gravity and Strings II. Proceedings, 1985/86. Edited by H.J. de Vega and N. Sánchez. V, 245 pages. 1987.

Vol. 281: Ph. Blanchard, Ph. Combe, W. Zheng, Mathematical and Physical Aspects of Stochastic Mechanics. VIII, 171 pages. 1987.

Vol. 282: F. Ehlotzky (Ed.), Fundamentals of Quantum Optics II. Proceedings, 1987. X, 289 pages. 1987.

Vol. 283: M. Yussouff (Ed.), Electronic Band Structure and Its Applications. Proceedings, 1986. VIII, 441 pages. 1987.

Vol. 284: D. Baeriswyl, M. Droz, A. Malaspinas, P. Martinoli (Eds.), Physics in Living Matter. Proceedings, 1986. V, 180 pages. 1987.

Vol. 285: T. Paszkiewicz (Ed.), Physics of Phonons. Proceedings, 1987. X, 486 pages. 1987.

Vol. 286: R. Alicki, K. Lendi, Quantum Dynamical Semigroups and Applications. VIII, 196 pages. 1987.

Vol. 287: W. Hillebrandt, R. Kuhfuß, E. Müller, J.W. Truran (Eds.), Nuclear Astrophysics. Proceedings. IX, 347 pages. 1987.

Vol. 288: J. Arbocz, M. Potier-Ferry, J. Singer, V.Tvergaard, Buckling and Post-Buckling. VII, 246 pages. 1987.

Vol. 289: N. Straumann, Klassische Mechanik. XV, 403 Seiten. 1987.

Vol. 290: K.T. Hecht, The Vector Coherent State Method and Its Application to Problems of Higher Symmetries. V, 154 pages. 1987.

Vol. 291: J.L. Linsky, R.E. Stencel (Eds.), Cool Stars, Stellar Systems, and the Sun. Proceedings, 1987. XIII, 537 pages. 1987.

Vol. 292: E.-H. Schröter, M. Schüssler (Eds.), Solar and Stellar Physics. Proceedings, 1987. V, 231 pages. 1987.

Vol. 293: Th. Dorfmüller, R. Pecora (Eds.), Rotational Dynamics of Small and Macromolecules. Proceedings, 1986. V, 249 pages. 1987.

Vol. 294: D. Berényi, G. Hock (Eds.), High-Energy Ion-Atom Collisions. Proceedings, 1987. VIII, 540 pages. 1988.

Vol. 295: P. Schmüser, Feynman-Graphen und Eichtheorien für Experimentalphysiker. VI, 217 Seiten. 1988.

Vol. 296: M. Month, S. Turner (Eds.), Frontiers of Particle Beams. XII, 700 pages. 1988.

Vol. 297: A. Lawrence (Ed.), Comets to Cosmology. X, 415 pages. 1988.

Vol. 298: M. Georgiev, F' Centers in Alkali Halides. XI, 287 pages. 1988.

Vol. 299: J.D. Buckmaster, T. Takeno (Eds.), Mathematical Modeling in Combustion Science. Proceedings, 1988. VI, 168 pages. 1988.

Turbulent Shear Flows

Editors: **L. J. S. Bradbury, F. Durst, B. E. Launder, J. L. Lumley, F. W. Schmidt, J. H. Whitelaw**

Volume 5

Selected Papers from the Fifth International Symposium on Turbulent Shear Flows, Cornell University, Ithaca, New York, USA, August 7-9, 1985

1987. 260 figures. VIII, 372 pages. ISBN 3-540-16885-0

This volume contains twenty-five papers presented at the Fifth Turbulent Shear Flows Symposium. The contributions have been considerably extended and updated to be of long-lasting value to the scientific community. They give a representative statement of current research in homogeneous and simple flows, free flows, wall flows and reacting flows. Each section begins with an introduction by a distinguished worker in the field. These surveys should be a great help not only to experts but also to newcomers to turbulent shear flow research. The reader will find coverage of the following topics: modelling of pressure terms of the scalar fluxes, inhomogeneous turbulence and applications to boundary layer flows, interactions of homogeneous turbulent scales, structural considerations, weak shear layer interactions, mixing layer flows and contribution flows are presented. The book should become a standard reference for physicists and engineers working in this exciting field of research.

Volume 4

Selected Papers from the Fourth International Symposium on Turbulent Shear Flows, University of Karlsruhe, Karlsruhe, FRG, September 12-14, 1983

1985. 286 figures. VIII, 397 pages. ISBN 3-540-13744-0

Contents: Fundamentals. - Free Flows. - Boundary Layers. - Reacting Flows. - Index of Contributors.

Volume 3

Selected Papers from the Third International Symposium on Turbulent Shear Flows, The University of California, Davis, September 9-11, 1981

1982. 244 figures. VIII, 321 pages. ISBN 3-540-11817-9

Contents: Wall Flows. - Scalar Transport. - Recirculating Flows. - Fundamentals. - Index of Contributors.

Volume 2

Selected Papers from the Second International Symposium on Turbulent Shear Flows, Imperial College London, July 2-4, 1979

1980. 310 figures, 12 tables. IX, 391 pages. ISBN 3-540-10067-9

Contents: Turbulence Models. - Wall Flows. - Complex Flows. - Coherent Structures. - Environmental Flows. - Index of Contributors.

Volume 1

Selected Papers from the First International Symposium on Turbulent Shear Flows, The Pennsylvania State University, University Park, Pennsylvania, USA, April 18-20, 1977

1979. 256 figures, 4 tables. VI, 415 Seiten. ISBN 3-540-09041-X

Contents: Free Flows. - Wall Flows. - Recirculating Flows. - Developments in Reynolds Stress Closures. - New Directions in Modeling.

G. Comte-Bellot, J. Mathieu, Ecully, France (Eds.)

Advances in Turbulence

Proceedings of the First European Turbulence Conference, Lyon, France, 1-4 July 1986

1987. 437 figures. XVI, 586 pages. ISBN 3-540-17586-5

Contents: Instability and Transitions. Theory and Experiments. - Chaotic Behaviour of Non-Linear Systems and Turbulent Fields. - Direct and Large Eddy Simulation of Turbulence. - Coupling of Fourier Modes. Spectral Analysis of Turbulence and Related Problems. - Two-Dimensional Velocity Fields. Geophysical and Astrophysical Turbulence. - Coherent Structures in Turbulence Flows. Conditional Averaging. Pattern Recognition. - Experimental Techniques. Hot-Wire Anemometry, Vorticity Meters, Electrochemical Methods, Image Analysis. - Engineering Applications of Turbulence and the Effect of External Disturbances. - List of Participants. - Index of Contributors.

Springer-Verlag Berlin Heidelberg New York London Paris Tokyo